The Political Ecology of Climate Change Adaptation

This book provides the first systematic critique of the concept of climate change adaptation within the field of international development. Drawing on a reworked political ecology framework, it argues that climate is not something 'out there' that we adapt to. Instead, it is part of the social and biophysical forces through which our lived environments are actively yet unevenly produced. From this original foundation, the book challenges us to rethink the concepts of climate change, vulnerability, resilience and adaptive capacity in transformed ways. With case studies drawn from Pakistan, India and Mongolia, it demonstrates concretely how climatic change emerges as a dynamic force in the ongoing transformation of contested rural landscapes. In crafting this synthesis, the book recalibrates the frameworks we use to envisage climatic change in the context of contemporary debates over development, livelihoods and poverty.

With its unique theoretical contribution and case study material, this book will appeal to researchers and students in environmental studies, sociology, geography, politics and development studies.

Marcus Taylor is an Associate Professor in the Department of Global Development Studies and the School of Environmental Studies at Queen's University, Canada.

Routledge Explorations in Development Studies

This Development Studies series features innovative and original research at the regional and global scale.

It promotes interdisciplinary scholarly works drawing on a wide spectrum of subject areas, in particular politics, health, economics, rural and urban studies, sociology, environment, anthropology and conflict studies.

Topics of particular interest are globalization; emerging powers; children and youth; cities; education; media and communication; technology development; and climate change.

In terms of theory and method, rather than basing itself on any orthodoxy, the series draws broadly on the tool kit of the social sciences in general, emphasizing comparison, the analysis of the structure and processes, and the application of qualitative and quantitative methods.

The Domestic Politics of Foreign Aid
Erik Lundsgaarde

Social Protection in Developing Countries
Reforming Systems
Katja Bender, Markus Kaltenborn and Christian Pfleiderer

Formal Peace and Informal War
Security and Development in Congo
Zoë Marriage

Technology Development Assistance for Agriculture
Putting Research into Use in Low Income Countries
Norman Clark, Andy Frost, Ian Maudlin and Andrew Ward

Statelessness and Citizenship
Camps and the Creation of Political Space
Victoria Redclift

Governance for Pro-Poor Urban Development
Lessons from Ghana
Franklin Obeng-Odoom

Nationalism, Law and Statelessness
Grand Illusions in the Horn of Africa
John R. Campbell

HIV and East Africa
Thirty Years in the Shadow of an Epidemic
Janet Seeley

Evaluation Methodologies for Aid in Conflict
Ole Winckler Andersen, Beate Bull and Megan Kennedy-Chouane

Digital Technologies for Democratic Governance in Latin America
Opportunities and Risks
Anita Breuer and Yanina Welp

Governance Reform in Africa
International and Domestic Pressures and Counter-Pressures
Jérôme Bachelard

Economic Development and Political Action in the Arab World
M. A. Mohamed Salih

Development and Welfare Policy in South Asia
Gabriele Koehler and Deepta Chopra

Confronting Land and Property Problems for Peace
Shinichi Takeuchi

Socio-Economic Insecurity in Emerging Economies
Building New Spaces
Khayaat Fakier and Ellen Ehmke

Foreign Aid and Emerging Powers
Asian Perspectives on Official Development Assistance
Iain Watson

The Political Ecology of Climate Change Adaptation
Livelihoods, Agrarian Change and the Conflicts of Development
Marcus Taylor

China's Foreign Relations and the Survival of Autocracies
Julia Bader

Democratic Accountability and Human Development
Regimes, Institutions and Resources
Kamran Ali Afzal and Mark Considine

"Embedding his narrative in powerful empirical studies of extreme-weather events in India, Pakistan, and the Mongolian steppes, Taylor produces the most incisive and sustained interrogation to date of the society/climate binary inherent in much that is written on climate-change adaptation. His own strategy of reading climate from a materialist point of view will no doubt provoke and enrich debates."

– Dipesh Chakrabarty, University of Chicago, USA

"For those suspicious of global calls for 'adapting' to climate change, Marcus Taylor provides ammunition and logic: an avalanche of detailed, intuitive, radical and compelling arguments and cases from around the world. For advocates of adaptation, he offers a grim and sobering reminder of the politically-loaded and careless violence of the international development machine."

– Paul Robbins, University of Wisconsin-Madison, USA

"Taylor's brilliant and pathbreaking new book explores the genealogy and construction of adaptation as a complex new field of knowledge and practice. It demonstrates how power, political economy and the production of vulnerability must be the foundations upon which new and radically transformative ideas and policies to combat climate change are constructed. A brave and important book."

– Michael Watts, University of California-Berkeley, USA

"This book provides a compelling answer for why it is that, although we know that climate change is a real and pressing issue, precious little real change is taking place. It offers an incisive analysis of adaptation and what might be wrong with it."

– Erik Swyngedouw, University of Manchester, UK

"Over the last two decades, climate change adaptation has increasingly emerged as the lodestar of public policy and development practice. Adapt now is the rallying cry of the moment. Taylor's brilliant and pathbreaking new book explores the genealogy and construction of adaptation as a complex new field of knowledge and practice. Rather than seeing adaptation as necessary and self-evident, Taylor exposes the ways in which the marriage of climate change and development rests upon conceptually untenable lines drawn between climate and society. He powerfully demonstrates how power, political economy and the production of vulnerability must be the foundations upon which new and radically transformative ideas and policies to combat climate change are constructed. A brave and important book."

– Marcus Taylor, University of California-Berkeley, USA

The Political Ecology of Climate Change Adaptation

Livelihoods, agrarian change and the conflicts of development

Marcus Taylor

Routledge
Taylor & Francis Group
LONDON AND NEW YORK

earthscan
from Routledge

First published 2015
by Routledge
2 Park Square, Milton Park, Abingdon, Oxon OX14 4RN

and by Routledge
711 Third Avenue, New York, NY 10017

Routledge is an imprint of the Taylor & Francis Group, an informa business

British Library Cataloguing in Publication Data
A catalogue record for this book is available from the British Library

Library of Congress Cataloging-in-Publication Data
Taylor, Marcus (Marcus Edward)
The political ecology of climate change adaptation : livelihoods, agrarian change and the conflicts of development / Marcus Taylor.
pages cm
1. Political ecology. 2. Climatic changes—Environmental aspects. 3. Political ecology—Pakistan. 4. Political ecology—India. 5. Political ecology—Mongolia. 6. Climatic changes—Environmental aspects—Pakistan. 7. Climatic changes—Environmental aspects—India. 8. Climatic changes—Environmental aspects—Mongolia. I. Title.
JA75.8.T38 2014
363.738′74095—dc23
2014019067

ISBN: 978-1-138-85913-5 (pbk)

Typeset in Goudy
by Apex CoVantage, LLC

Contents

Illustrations

Figures

Tables

Box

Acknowledgements

In writing this book I have accrued many debts. In the institutional realm, the research was supported through a Social Sciences and Humanities Research Council grant on 'Climate Change and Social Conflict in the Himalayan Watershed'. A version of Chapter Six was previously published as "Liquid Debts: credit, groundwater and the social ecology of agrarian distress in Andhra Pradesh, India" in *Third World Quarterly*, 34:4, 691–709, 2013. Parts of Chapter Four were published as "Climate Change, Relational Vulnerability and Human Security: Rethinking Sustainable Adaptation in Agrarian Environments" in *Climate and Development*, 5(4), 318–327, 2013. The publishers have kindly granted permission to reproduce that material.

A number of people have been instrumental in aiding the project from its inception. T. Jayaraman was extremely kind to host me at the Tata Institute of Social Sciences in Mumbai in the winter of 2012, and Suhas Palshikar likewise invited me to present at the University of Pune. I gained considerable insight through the discussions, both formal and informal, that I had with colleagues from both these faculties. I also benefited greatly from the 2012 workshop on debt held at Queen's University, with particular thanks to the organisers Adrienne Roberts and Susanne Soederberg and the other participants for their feedback. At Queen's University, Anoosh Noorizadeh Kollou aided the project greatly with his research assistance on Pakistan, while Keira Loukes provided similarly excellent assistance for the chapter on Mongolia. Mahmood Khan kindly sent me a copy of his most recent book on Pakistani agriculture. Bethany Wright and Charlotte Russell both aided the book – with patience – through the publication process.

A range of colleagues kindly read and commented upon select draft chapters, including Jayant Lele, Haroon Akram Lodhi, Philip McMichael, Mark Carey, Dia da Costa, Dean Curran, Keira Lourkes and Stacey Murie. In particular, I must thank most fully Susanne Soederberg and Sébastien Rioux, both of whom read through a range of chapters and provided detailed and most helpful suggestions. They are, of course, not to blame for my many remaining errors. I also thank Jayant and Dorothy, Dia and Alex, Martin and Mandy, Mick and Joyce and my parents Anne and Barry for various types of support during the project.

Most importantly, Susanne, Sydney and Teivo provided a wonderful source of companionship and fun at home. It is to them that I owe the very biggest debt of gratitude and to whom I dedicate this book with much love!

Preface

The critique of climate change adaptation

This book interrogates the emergence of climate change adaptation as a new and complex field of knowledge production and development practice. With a specific focus on agrarian regions, my entrance point into the issue is through a close analysis of the discourses and policies associated with national governments and international development agencies whose actions are commonly packaged under the rubric of development. Climate change, it is roundly acknowledged, greatly complicates both present practices and future expectations within this field. The United Nations Development Programme, for example, labels climate change as the "defining human development issue of our generation" and one that challenges the enlightenment aspiration of a collective journey of humanity towards a better future (UNDP 2007: 1). Such concerns stem from the overwhelming consensus within scientific and development organisations that global climate change is triggering profound transformations in social and ecological systems that will cause significant dislocations and stress among affected populations (IPCC 2007). The most severe impacts, moreover, are commonly projected to be concentrated among the world's poor and particularly those living in rural areas of the global South (World Bank 2010).

Given the severity and unequal distribution of projected climate change impacts, international institutions and national governments have advanced the pressing need for rapid and far-reaching processes of climate change adaptation. In normative terms, climate change adaptation is described as a process of transformation in social and environmental systems that can safeguard against current and future adverse impacts of climatic change. Simultaneously, it is also envisioned as a process that facilitates societies to take advantage of any new opportunities provided by a changing environment (IPCC 2007; World Bank 2010). In practice, while the goal of adaptation might be realised through the spontaneous and unstructured behavioural alterations by individuals and social groups – such as farmers changing crops, households diversifying livelihoods, families migrating from exposed regions – such 'autonomous adaptation' is imagined to be insufficiently encompassing to deal with the gravity of projected threats. Adaptation, therefore, is viewed predominantly as a process of coordinated transition to meet the demands and challenges of a changing external environment directed by appropriate governmental institutions (United Nations Framework

Convention on Climate Change 2007). It is on this basis that measures to address climate change are argued to require immediate mainstreaming within both national policymaking and international development initiatives. Facilitating climate change adaptation, it seems, has become a litmus test for the project of development.

In response, a burgeoning academic and policy literature has emerged to help meet this aim. This literature is broad and, as is set out in the following chapters, different perspectives within the field debate the appropriate sites and scales of adaptation, the rights and responsibilities of affected and contributor groups and the necessary mechanisms and goals of adjustment (Pelling 2011). Although this body of work is diverse, and occasionally fractious, it is bound together by the shared assumption of a common and collective need to adapt. "Adaptation now!" has become a shared refrain of international institutions, national governments, non-governmental organisations and scholars working in the field (e.g. Adger et al. 2009; Leary et al. 2010). The idea of adaptation, therefore, has become a touchstone concept that provides both a normative goal and a framework within which practical interventions are planned, organised and legitimised. Rapidly incorporated into the governmental lexicon of development, the idea of adaptation circulates as the accepted rubric for conceptualising social transformations under anthropogenic climate change. From the paddy fields of Uttar Pradesh to the growing shantytowns of Ulaanbaatar, the collective threat stemming from climatic change has seemingly propelled us into a common yet uneven world of adaptation.

In this rush to marry climate change adaptation and development, however, there remains relatively little critical enquiry into the idea of adaptation that underpins such governmental energies. In part, this is because adaptation is commonly cast as a natural moment of transformation that reflects a process common to all forms of life. From its roots in evolutionary biology, adaptation projects the necessity for organisms to constantly adjust to changes in their external environment as a means to bring themselves in line with new constraints and opportunities. Extracted from its roots in biology and transposed into the context of contemporary climate change, adaptation is now held to represent an equally innate process of social adjustments to external climatic stimuli. Facing the assuredly grave consequences of global climatic change, the pressing need for immediate and comprehensive adaptation is seemingly self-evident. As Adger, O'Brien and Lorenzoni put it, "we already know that adaptation is necessary" (2010: 2).

Over the following chapters, however, I set out the argument that we should be exceedingly wary of such representations. To this end, the book interrogates climate change adaptation not as a self-evident analytical framework and normative goal but as an array of discursive coordinates and institutional practices that themselves form the object of analysis. To do so, I pay close attention to the ways that the concept of adaptation fashions a relatively cohesive body of ideas around the relationship between climate change and society into which issues of social change, power and environmental flux are placed and solutions drawn.

At its core, the adaptation framework is predicated upon an inherent dichotomy between climate and society in which the former is represented as a cohesive external system that generates threats, stresses and disturbances, and the latter is portrayed as a separate domain of social structures that are unevenly vulnerable to climatic change. Through this representational regime the discourse produces its 'world of adaptation' in which all social units can be understood and acted upon in terms of a universal schematic of exposure to external climatic threats. The idea of adaptation thereby consolidates a social imaginary of individuals, households, communities, regions, economic sectors and nations with different vulnerabilities and adaptive capacities in the face of an external climate that, tipped off balance by the unintended actions of humans, is dangerously off-kilter.

Through this imagery of climate as an external threat that renders regions and people vulnerable to its capricious nature, the adaptation framework is remarkably successful in creating a new object for development interventions. A world of adaptation can be mapped out in terms of a social cartography of vulnerabilities to be ameliorated by building adaptive capacity and forging resilience. This intrinsically biopolitical impetus to make climate change governable, however, comes at the expense of obscuring crucial political questions about power and sustainability within the ongoing production of our lived environments. The idea of adaptation, I argue, intrinsically lends itself to a technocratic politics that seeks to contain the perceived threats posed by climate change within existing institutional parameters. On this basis, I argue that the seeming naturalness of adaptation stands as a considerable barrier to critical thinking about climatic change and social transformation.

There are, of course, a number of contributions to the adaptation literature that are pointedly critical of a technocratic reading of adaptation. Karen O'Brien and collaborators pointedly ask what is at stake in different framings of vulnerability within the adaptation framework (O'Brien et al. 2007). They argue that while scientific framings produce a managerial discourse that privileges technological solutions to adaptation, a human-security framing builds from the question why some groups and regions are more vulnerable than others, therein facilitating a different politics of adaptation. The purpose of such interventions is to make adaptation more attuned to the needs of the poor and marginalised who are faced with the double burden of existing inequalities coupled with greater risks from climatic change (Eriksen and O'Brien 2007; St. Clair 2010; see also, Brown 2011).

There is a considerable amount of important and instructive work undertaken within this approach and the following chapters undeniably demonstrate its keen influence. It is striking, however, that even this critical counterpoint maintains adaptation as a given and self-evident concept. Although such perspectives rightfully emphasise how social marginalisation and inequality unevenly stratify the impacts of climate-related shocks, they continue to operate within adaptation's schematic of external climatic threats and internal social exposures. As such, they maintain the framework of adaptation but seek to leverage policymaking in a progressive and transformative direction (O'Brien, St. Clair and Kristoffersen 2010; Pelling 2011). What they do not offer, however, is a critique that questions the very notion of 'adaptation' as a prima facie category of analysis and practice. To do so is to de-frame climate change adaptation to render visible its embedded assumptions and contradictions. Instead of accepting adaptation as a self-evident concept, therefore, the present book deconstructs it as a framing device that profoundly limits how we conceptualise climatic change, its impacts and our potential responses.

The analytical core of this intervention is set out in the first three chapters, in which I critique the Cartesian foundations of adaptation that dichotomise climate and society as two separate yet mutually influencing systems or domains. It is this dichotomy, I argue, that leads towards the representation of climate change as a series of external shocks and disturbances to an otherwise coherent society. Through this separation, climate change is parsed out and isolated from the ongoing processes of social and ecological transformation that construct our lived environments. The purpose of such a separation is precisely to make climate change governable as a managed process of adaptation. What is missing in such representations, however, is that humans do not stand outside their environments but are active protagonists in their ongoing production. As the presently fashionable concept of the anthropocene indicates, this intrinsically involves the production of climate. Under such conditions, the adaptation framework of distinct yet interacting natural and social systems seems curiously unsuited to a world in which what we term 'nature' has become increasingly produced through human activities.

From this perspective, the Cartesian dichotomy between climate and society as separate and external domains that undergirds the adaptation framework is rendered problematic. Instead of conceptualising climate and society as bounded entities, wherein one influences, impacts or threatens the other, the book develops the concept of 'material climates' in which social and meteorological dynamics are seen as fundamentally intertwined, co-productive, constantly refashioned and changing. Rethinking the concept of climate impels us to explore climatic change in terms of the shifting couplings of human and meteorological forces through which our lived environments are actively formed. Political ecology is central to this task because its analytical tools help us capture how meteorological processes are embedded within hierarchically ordered social relationships in ways that produce strikingly uneven and often deeply fragile landscapes. From this perspective, we can approach climate change not as a rupture between society and a climate thrown out of balance by human actions but as a series of tensions in the way that meteorological forces are actively worked into the production of our lived environments. On this basis, climate change represents a shift in the socio-ecological relationships through which our lived environments, with all their engrained inequities and forms of power, are actively produced. The political implication is worth highlighting: instead of converging on the imperative to adapt, we must instead focus on producing ourselves differently.

To concretise this intervention, I turn to a close empirical examination of agrarian environments in South and Central Asia. In these contexts, I argue that the framework of climate change adaptation has emerged as a new and intrinsically political domain of development practice that operates within a wider spectrum of governmental technologies that represent, order and reshape the agrarian world. To understand how adaptation operates as a governmental practice it is necessary to place the contemporary experience of climatic change within a longer historical register of social and environmental transformation in which agrarian spaces have stood at the nexus of conflicting designs and agencies. It is only by situating adaptation within this broader terrain of agrarian transformation that we can come to terms with the political dimensions of what it means to understand climate change through the framework of adaptation. On this basis, the book demonstrates how the rhetoric and practices of adaptation operate within a deeply political terrain that is configured by contested normative visions of agrarian space. The latter emerge in the context of diverse projects aimed at recalibrating rural regions driven by governments, institutions, corporations and social movements. Climate change adaptation, therefore, is intrinsically a political process despite its pretensions otherwise.

To this end, Chapters Four and Five read climatic change through the lens of political ecology to pose questions that are rarely touched upon within the adaptation literature. They ask what it entails to set climate change impacts within agrarian regions in which the production of goods has been relentlessly commodified, drawing both human livelihoods and their socio-ecological foundations into circuits of capital accumulation that operate on scales that far exceed the specific locality of production. Simultaneously, they ask how we should understand the concepts of vulnerability and resilience in the context of ongoing processes in which rural labour forces are being decomposed and recomposed in new and complex forms. To pose these questions is to situate what is termed 'adaptation to climate change' as part of wider historical processes of agrarian transformation and forms of governmentality within rural regions (see, Davis 2002). Although such concerns are anathema to the adaptation discourse, the book stresses the need to understand climate change in the context of the uneven commercialisation of agriculture, changing property relations, forms of capital accumulation, the dynamics of state formation, macro-projects of environmental engineering, migratory flows, technological change and the emergence of new rural subjectivities and political movements.

By bringing such contested trajectories into the heart of our analysis, we can begin to understand how contemporary climatic change interacts with these dynamics in new and complex ways. Through this analysis, climate change emerges as part of ongoing historical processes of socio-ecological transformation predicated upon forms of power operating at varied spatial scales that shape control over land, water, bodies and debt. As I map out in Chapter Four, these socio-ecological relations construct the parameters through which households seek to reproduce themselves by distributing the essential insecurities of agrarian life in a relational and hierarchical manner. To think of vulnerability in relational terms is to uncover the socio-ecological relations through which the security of some and the relative insecurity of others are directly intertwined. Although it falls out of the analytical purview of the adaptation framework, this relational focus forms a core aspect of my political ecology analysis. It emphasises that what the discourse labels 'adaptation to climate change' is fundamentally rooted in questions of power and production.

This framework is then used to analyse three case studies that situate questions of climatic change and agrarian transformation across distinct socio-ecological settings and historical contexts. The first study historicises the discourse of climate change adaptation arising within the devastating floods of 2010–2011 that impacted much of rural Pakistan. It does so by demonstrating how the localistic and presentist frames that dominate the adaptation literature obscure the longer trajectories of agrarian transformation in the region. In tracing the political ecology of agrarian relations from the colonial period onwards, the chapter explores the long-term construction and reproduction of vulnerability within the changing contours of ecological change and the shifting incorporation of agricultural production into world markets. It demonstrates how repeated attempts to engineer the socio-ecology of the Indus watershed since colonial times are intrinsic to the contradictory dynamics of agrarian transformation occurring in the present. This provides the basis for a close critique of the technocratic and managerial rendering of adaptation adopted by the Pakistani government. Notably, the question of land redistribution emerges as a key strategy for transforming rural Pakistan within the context of climatic change, despite its complete marginalisation in both government approaches to the issue and the adaptation paradigm in general.

Moving to a regional level, the second study examines relationships of debt and vulnerability in the semi-arid Deccan plateau in southern India. In the context of the increasing frequency of drought, the chapter examines the intersection of climate variability, enduring debt relations and uneven access to water in conditions of an agrarian environment transformed by the liberalisation of agricultural policy. The deleterious impacts of climatic change upon agricultural production in this region are situated within the context of an agrarian environment already haunted by unprecedented numbers of farmer suicides. The chapter details how the agrarian dynamics of contemporary semi-arid Andhra Pradesh are strongly determined by the tenacious yet highly tenuous attempt to secure social reproduction undertaken by a large class of marginal and smallholder farmers that precariously struggle to carve out livelihoods. In this context, the control over water and credit form inseparable parts of the socio-ecology of agrarian transformation under complex capitalist dynamics. The uneven access to credit for well drilling became central to gaining control over the irrigation necessary for increasingly specialised commercial agriculture in conditions of liberalisation and new technologies. At the same time, endemic debts drive on the risks of agricultural failure in the context of rapidly depleted shallow aquifers that characterise the Deccan regions of central and southern India. This intersection of climatic change, fickle waters and enduring debts not only configured a new nexus of insecurity for smallholders but also became integral to the dynamics of surplus extraction and the unequal distribution of risk across the agrarian environment. This raises pressing political questions around smallholder agriculture that are entirely marginalised within the confines of the adaptation paradigm.

The third case examines the political ecology of the Mongolian steppe, where pastoral livelihoods are argued to be uniquely vulnerable to climatic change. In this context, there have been repeated calls to improve environmental and cultural conservation and build community resilience as a means to adaptation. Obscured in such narratives, however, is how successive structural adjustment programmes placed immense strain upon the herding economy through deindustrialisation and the imposition of a changing property regime over the grasslands, leading to increased herd sizes and a tendency towards overgrazing. These dynamics led to a crisis of the pastoral economy that was brutally exposed as a succession of extremely cold winter storms (*dzuds*) destroyed herds. Presently, the pastoral economy faces not only these socio-ecological contradictions but also the dramatic expansion of mining. As part of a new frontier of capital accumulation based on intensive resource extraction, Mongolia is estimated to have enough coal to fire every power station in China for the next 50 years. The irony here is that such coal-fired energy production is contributing to the climatic change at both regional and global scales that further undermines pastoral livelihoods. Interrogating these sharp tensions emphasises how the future of Mongolian pastoralism is shaped within global flows of finance, energy, raw materials and pollutants that are largely excluded from the discourse of climate change adaptation.

These cases impel us to address climate change outside the terms of adaptation so as to widen our political horizons. As the book notes, confronting climate change is not about adapting to an external threat. Instead, it is fundamentally about producing ourselves differently. In moving beyond the adaptation paradigm, two central political questions emerge. First, we need to explicitly foreground ways to collectively deleverage a global capitalist order that is predicated upon the unending accumulation of productive forces and consumptive practices that give rise to the deadly metabolisms inherent to climatic change. This requires opening up the fundamental premises of development and its teleology of globalising boundless consumption. Second, it raises the need to reimagine redistribution as a central pillar of future equitable socio-ecological transformation. Within agrarian environments, redistributive strategies – from land and water rights to credit policies and subsidies – historically have been a central aim of many agrarian social movements. Despite their marginalisation within the framework of climate change adaptation, these struggles become ever more important within the context of contemporary climatic change. Indeed, the inherent and widely recognised inequities of climatic change potentially open a pathway towards revitalising the idea of redistribution across spatial scales. Thinking beyond adaptation will be central to turning such possibilities into practice.

References

Adger, W. N., S. Dessai, M. Goulden, M. Hulme, I. Lorenzoni, D. Nelson, L. O. Naess, J. Wolf, and A. Wreford. 2009. "Are there social limits to adaptation to climate change." *Climatic Change* 93(2): 335–54.

Adger, W. N., I. Lorenzoni, and K. O'Brien. 2010. "Adaptation Now." Pp. 1–22 in *Adapting to Climate Change*, edited by W. Neil Adger, I. Lorenzoni, and K. O'Brien. Cambridge: Cambridge University Press.

Brown, K. 2011. "Sustainable adaptation: An oxymoron?" *Climate and Development* 3(1): 21–31.

Davis, M. 2002. *Late Victorian Holocausts*. London: Verso.

Eriksen, S., and K. O'Brien. 2007. "Vulnerability, poverty and the need for sustainable adaptation measures." *Climate Policy* 7(4): 337–52.

IPCC. 2007. *Climate Change 2007: The Scientific Basis. WG I Contribution to IPCC 4th Assessment Report*. Cambridge: Cambridge University Press.

Leary, N., J. Adejuwon, V. Barros, P. Batima, B. Biagini, I. Burton, . . . M. Wehbe. 2010. "A Stitch in Time: General Lessons from Specific Cases." Pp. 1–27 in *Climate Change and Adaptation*, edited by Neil Leary et al. London: Earthscan Press.

O'Brien, K., S. Eriksen, L. Nygaard, and A. Schojolden. 2007. "Why different interpretations of vulnerability matter in climate change discourses." *Climate Policy* 7(1): 73–88.

O'Brien, K., A. L. St. Clair, and B. Kristoffersen (Eds.). 2010. *Climate Change, Ethics and Human Security*. Cambridge: Cambridge University Press.

Pelling, M. 2011. *Adaptation to Climate Change*. London: Routledge Press.

St. Clair, A. L. 2010. "Global Poverty and Climate Change: Towards the Responsibility to Protect." Pp. 180–98 in *Climate change, ethics and human security*, edited by Karen O'Brien, Asuncion Lera St. Clair, and Berit Kristoffersen. Cambridge: Cambridge University Press.

UNDP. 2007. *Human Development Report 2007/2008: Fighting Climate Change – Human Solidarity in a Divided World*. Geneva: United Nations.

United Nations Framework Convention on Climate Change. 2007. *Climate Change: Impacts, Vulnerabilities and Adaptation in Developing Countries*. Bonn: United Nations.

World Bank. 2010. *The Economics of Adaptation to Climate Change*. Washington, DC: IBRD.

1 Climate change and the frontiers of political ecology

Since the advent of historical capitalism, virtually no part of the planet has remained untouched by humanity's restless compulsion to transform nature. It is now more than a century and a half ago that Marx and Engels wrote effusively about humanity's newly awakened productive powers that cleared "whole continents for cultivation" and simultaneously conjured "entire populations out of the ground" (Marx and Engels 1998). Their arguments reflected the degree to which humans had become prolific agents of environmental change on a world scale, therein anticipating what some authors now term 'the anthropocene' (Crutzen and Steffen 2003). This Promethean project of harnessing nature to anthropogenic designs appeared to be the realisation of modernity's founding premise that humans could collectively create and enact their own future outside of determination by natural laws. Such ethos, however, held a dark underside. The pursuit of rationality, efficiency and accumulation on a global scale travelled hand-in-hand with the historical processes of enclosure, expropriation, domination and enslavement (Wolf 1982). Moreover, while the unleashing of humanity's productive energies created a world of unparalleled – if desperately unequal – consumption, it also left a trail of resource depletion, land degradation, environmental pollution and species extinction (UNEP 2014). Attempting to mediate or reverse such contradictory forces has been the source of intense and bitter social struggles across the history of world capitalism (Gadgil and Guha 1993; Grove 1997; Martínez-Alier 2002).

Contemporary climate change, however, appears to pose a different order of questions. Whereas the use and abuse of nature noted above encountered notable biophysical constraints, these often appeared to be relatively localised and permeable limits to human designs. Within capitalism, as Marx noted, every limit appears as a barrier to be overcome and the ensuing history of capitalism is one of compulsive technological change, the opening of new resource frontiers and the repeated displacement of such 'externalities' onto the human and geographical margins of society (Marx 1973: 408; Moore 2010a; Barbier 2011). The idea of anthropogenic climate change, however, appears to level a much greater challenge to embedded modernist convictions and practices. Here, nature manifests itself not as a passive resource that strains and complains under human demands but as a dynamic historical agent with the potential to dramatically shape humanity's future on a planetary scale. As David Clark provocatively notes, the current suspicion that humankind has turned the planet's weather systems into a vast experiment has an ominous supplement: the recognition that drastic climatic shifts have experimented with human life across history in ways that have repeatedly put humans through desperate trials and hardships (2010: 32). On these grounds, by collectively releasing vast amounts of sequestered carbon into the atmosphere, humanity's agency is conceived to have awoken a dangerous leviathan from its brief geological slumber with uncertain historic consequences (Fagan 2004).

Under the spectre of rapid and profound climate change, a new social topography of risk has emerged. Humanity's relationship to nature no longer appears as a domain of controlled manipulation. Instead, it opens a fissured terrain of profound vulnerability scoured by the power of capricious climatic forces. Such inversions have inevitably created profound anxieties concerning humanity's ability to shape its own future (Chakrabarty 2009; Hulme 2010). According to the United Nations Development Programme (UNDP), climate change calls into question the very ideas of development and progress to which the project of modernity is tethered. Failure to recognise and deal with the effects of climate change, it estimates, will consign the poorest 40 per cent of the world's population to a future of diminished opportunity and will sharpen the already acute divisions between the 'haves' and 'have-nots' (UNDP 2007). On these grounds, climate represents a powerful agent of anti-development that, left unchecked, will roll back the already uneven achievements of the modern era.

In response, a dominant policy and academic literature has hastily emerged under the banner of climate change adaptation. This body of work builds from the seemingly self-evident proposition that, if the climate is changing in ways that threaten the existing parameters and future well-being of society, humanity must adapt through a process of planned adjustment that can safeguard against such profound and escalating risks (IPCC 2007). The idea of adaptation has therein become a rallying cry intended to catalyse a determined human response to the threats posed by climate change (Adger, Lorenzoni and O'Brien 2010; Leary et al. 2010). Considerable governmental energies are currently leveraged in its pursuit. Noticeably, in the field of international development, the goal of climate change adaptation now acts as a shared rubric for a diversity of planned interventions, drawing international agencies, governments, corporations, non-governmental organisations (NGOs) and social movements into a common and encompassing framework (Ireland 2012).

Notwithstanding a great deal of sympathy with the stated intentions of adaptation as a normative goal, in what follows I argue that its framework should not be considered an exclusive way of conceptualising the acute challenges that climatic change duly raises. On the contrary, despite its current dominance in academic and policy debates, the salience of adaptation within contemporary policymaking rests less on its conceptual integrity and more on its ability to render climatic change legible to the registers of governmental planning. This intrinsically biopolitical impetus, I contend, comes at the expense of obscuring vital political questions surrounding power and sustainability in an era of dynamic global transformations. Rather than proceeding from the foundation of adaptation, this book asks instead how we might read contemporary climate change differently through the lens of political ecology. While I do not provide a systematic reconstruction of political ecology as a field – a task which has been variously undertaken elsewhere (e.g. Peet and Watts 2004; Neumann 2005; Robbins 2012) – I seek here to illustrate its compelling features as an entry point into analysing the narratives and practices through which climate change is both produced and experienced.

To do so, the chapter draws together a series of shared concerns about power, representation and the production of lived environments that bind political ecology together as an analytical framework. First, I take seriously the notion of political ecology as a field that duly combines the concerns of ecology and political economy in a way that "encompasses the constantly shifting dialectic between society and land-based resources, and also between classes and groups within society itself" (Blaikie and Brookfield 1987: 17). I elaborate how this perspective allows us to get to the core of the relational dimensions of a global political ecology in which the couplings of prosperity and marginalisation, security and vulnerability, and abundance and degradation are produced and reproduced together through overlapping structures of power across spatial scales (Blaikie et al. 1994; Peet, Robbins and Watts 2011b). Subsequently, the chapter engages with a second pillar of political ecology analysis that considers how representation forms an inherent dimension of such power relations (Escobar 1995; Peet and Watts 1996; Escobar 1999; Blaikie 2001). Following this trajectory, I chart the ways in which climate change adaptation operates as a discursive apparatus that renders climate change legible in a narrow and constrained fashion. In particular, I critique its grounding notion of climate as an external system that provides exogenous stimulus and shocks to which society must then adapt. The latter dichotomy, I note, appears peculiarly unsuited to a world in which human and meteorological forces have become intrinsically intertwined and co-productive.

To go beyond the imagery of society and climate as separate systems locked into an endless dance of adaptation, I argue that we must push at the frontiers of political ecology by drawing insights from radical geography (Smith 1984; Harvey 1996; Castree 2001), urban political ecology (Swyngedouw and Heynen 2003; Swyngedouw 2004; Kaika 2005), poststructuralist 'more-than-human' ontologies (Latour 1993; Bennett 2010; Head and Gibson 2012) and ecological anthropology (Ingold 2000, 2011). In so doing, the chapter draws out how a reworked political ecology framework can help us grapple with the complex couplings of human and meteorological forces through which our lived environments are actively yet unequally produced. This approach, I contend, provides a means by which we can write questions of power more articulately into our analyses of climate change and social transformation. It, therefore, opens a deeper set of political questions about power, production and environmental change than is possible within the paradigm of climate change adaptation.

Political ecology and the critique of adaptation

For many analysts grounded in the early works of political ecology there likely arises a sense of déjà vu when surveying the current debates on climate change adaptation. A sharp engagement with the paradigm of cultural ecology and its core concepts of adaptation and homeostasis was one of the birthing grounds of political ecology as a field in the 1980s. For cultural ecologists, the concept of adaptation provided an analytical framework by which to situate the relative ability of humans to respond flexibly to shifts in their environment as part of a broader processes of human cultural evolution (Harrison 1993). From climatic shifts to land degradation, humans were seen to react to environmental change by first coping with and then adapting to successive series of external stresses and stimuli. This ongoing process of adaptation, however, required changes not only the way that humans engaged with the natural environment – such as shifts in cropping or migrations to exploit new ecological niches – but also in the belief systems that structured such practices. For cultural ecologists, therefore, the concept of adaptation described a cumulative series of adjustments comprising the interaction of social practices, systems of meaning and technological changes that might enhance the ability of a given community to cope with environmental stresses (Rappaport 1979). The expected result of such adaptive strategies was not simply a process of behavioural change but rather of a broader cultural evolution that could realign human activities and belief systems with the demands of a changing external environment. Successful adaptation, therefore, created the grounds for a new homeostasis or equilibrium in the relationship between communities and their natural environments.

For early political ecologists, both the analytical framework and political conclusions of adaptation analysis appeared to be problematic. In proposing the centrality of engrained belief systems to homeostasis, the explicit functionalism of adaptation analyses easily could be inverted to frame environmental degradation as the outcome of entrenched yet irrational forms of land management resulting from traditional values that were rendered anachronistic in a rapidly changing world (Blaikie 1985; Blaikie and Brookfield 1987; see also, Robbins 2012). As such, although cultural ecologists often celebrated the lifestyles of the farming, hunting and herding groups they studied, the narrative of adaptation could be reworked for quite different purposes. For modernisation theorists, the demands of economic development required a profound transformation in the value orientations of postcolonial agrarian populations to overcome their perceived proclivity for subsistence-orientated and risk-adverse livelihoods. The political stakes were high. Under the lens of modernisation, a failure to crack the nut of traditional agricultural practices and their associated belief systems could leave societies trapped in a stagnant dynamic in which resource use would remain inefficient and prone to depletion under the pressures of population growth. Authors such as Bert Hoselitz, therefore, were remarkably brazen about what must be done:

> Value systems offer special resistances to change, but without wishing to be dogmatic, I believe, it may be stated that their change is facilitated if the

> material economic environment in which they can flourish is destroyed or weakened. This seems to be the experience from the history of Western European economic development, and it seems to be confirmed by the findings of students of colonial policy and administration.
>
> (Hoselitz 1952: 15)

For political ecologists, the political ambivalence of cultural ecology's adaptation analysis stemmed from its marginalisation of a crucial set of historical dynamics that were busily shaping agrarian environments. In contrast to the self-regulating localism of adaptation perspectives, political ecologists sought to situate localised processes within a multi-scalar series of causal forces. Far from isolated regions of untouched tradition, authors such as Piers Blaikie (1985) emphasised how rural regions betrayed the complex outcomes of colonial forms of land management and incorporation into capitalist commodity relations, both of which had diverse and contradictory effects upon local social relations and environmental landscapes. In this reading, the problems of land degradation, soil erosion and deforestation could not be placed at the feet of 'irrational' peasants who failed to adequately adapt to changing social and environmental stimuli. Rather, those biophysical trends spoke to the way that integrating agricultural production within regional and international accumulation dynamics created new forms of enclosure and surplus extraction that disrupted the socio-ecological fabric of rural regions (Blaikie, Cameron and Seddon 1983; Blaikie 1985; Blaikie and Brookfield 1987). In transforming agrarian environments and producing new forms of marginality, these social forces created the grounds upon which peasants were increasingly pressured to act as agents of environmental degradation in a fraught struggle to meet subsistence needs (Watts 1983).

This analytical perspective posed a direct challenge to the narrow conceptual framework of adaptation. As Richard Peet and Michael Watts put it, "market integration, commercialisation and the dislocation of customary forms of resource management – in place of adaptation or homeostasis – became the lodestars of a critical alternative to the older cultural or human ecology" (2004: 9). Conspicuously, the emphasis on social differentiation under the forces of capitalist commodity production allowed political ecology to question who or what could be said to 'adapt'. While cultural ecology tended to represent rural communities as relatively cohesive and bounded entities, political ecologists argued that such representations obscured the fractured social terrain of rapidly changing agrarian spaces and the diversity of competing interests within them (Robbins 2000). In so doing, political ecologists tugged at the analytical seams of the adaptation concept in a way that still holds resonance for contemporary debates. In place of unitary communities struggling to adapt to external stresses, political ecology emphasised how hierarchical forms of local resource management were consolidated under power differentials built upon relations of class, gender, caste and ethnicity (Mosse 2007). Such fractures, moreover, also reflected the divergent ways that social groups were situated within networks of commodity production and institutionalised political power that stretched far outside the locality in question (Watts 2004). What could adaptation signify in conditions where social groups experience the gains and risks inherent to social and ecological transformations in profoundly different and unequal ways? As such, once the image of a homogenous community with a relatively unitary set of interests was rejected, the idea of adaptation appeared less a valid analytical tool than a politically constituted concept liable to smother over the social fractures that permeated agrarian regions (Leach, Mearns and Scoones 1999).

From cultural ecology to climate change adaptation

Although the popularity of adaptation as a core analytical concept flagged in the 1990s, in part due to the concerted critiques levied by early political ecologists, the emergence of climate change as a core domain of governmental concern in the new millennium has led to its dramatic revival (Head 2010; Pelling 2011; Bassett and Fogelman 2013). Adaptation to climate change, as the commonly used definition states, is the adjustment in natural or human systems in response to actual or expected climatic stimuli or their effects that moderates harm or exploits beneficial opportunities (IPCC 2007). While this emphasis on adaptation being a process of adjustment to climatic shocks unifies the literature, different frameworks provide distinct answers to key underlying questions. They vary on the questions of who or what is to adapt? How are they to do so? And what are the ends of adaptation? As such, distinct traditions within the paradigm of climate change adaptation incorporate different ideas of the appropriate sites and scales of adaptation, the rights and responsibilities of affected and contributor groups and the necessary mechanisms and goals of adjustment. Consequently, they legitimate different policy responses and forms of intervention (see Chapters Three and Four).

The current usage of adaptation within the climate change literature, therefore, is significantly broader and more diverse than that of cultural ecology. Viewed from the perspective of the cumulative body of work within the political ecology tradition, it nonetheless appears to share several of the latter's weaknesses. First, there is a frequent tendency to conceive of regions and landscapes affected by climate change as bounded domains upon which climatic stresses emerge as a new and externally generated threat. This framework is captured in the systems language of adaptation noted above and leads to what Michael Watts cautioned was a billiard ball view of the world in which pre-constituted entities collide to cause change (1983). Through an imagery of regions facing the approaching eight-ball of climate change, this perspective tends towards an examination of vulnerability in a synchronic manner that conceives of vulnerability in terms of exposures to an external threat. Regardless of whether we consider such vulnerability to be determined more by the properties of the external shock (e.g. the magnitude of a cyclone or the length of a drought) or the level of internal exposure (e.g. the presence of social inequalities, a lack of institutional capacity), it retains a model predicated upon a relatively static inside/outside dichotomy. This orientates analysis towards a perspective that is strongly bound by localism and presentism – what I term the 'here and now' of adaptation – in which vulnerability is conceived as an anomalous condition to be identified, intervened upon and resolved, thereby paving the way for managerial and technocratic interventions. On this basis, it is not surprising that Bassett and Fogelman's (2013) extensive survey of the adaptation literature showed that over 70 per cent of academic publications on the subject presented adaptation as a technical process of planned social engineering to guard against proximate climatic threats.

Second, yet stemming from the former, there remains a pervasive reluctance within the current adaptation literature to conceptualise the varied forms of power that shape how different social groups are rendered secure or vulnerable to environmental change. As a consequence, while the field stands awash with claims about vulnerable peoples and stressed ecosystems facing an external threat, it is conspicuously short of the kind of historically and analytically grounded analyses that shed light on the situated socio-ecological relations that produce social vulnerability and environmental fragility. Notably, in the context of agrarian environments, climate change adaptation is repeatedly represented as a case of adjusting regions and communities to climatic threats with scant attention paid to the historical roots of the vulnerability that many marginal groups face. In this respect, there emerges an unnerving sense that the literature consistently sidesteps core questions concerning the historically shaped and hierarchically ordered control over land, water, capital and labour that typically characterises rural regions and unequally distributes risks and rewards within them (e.g. World Bank 2008; IFAD 2010, 2013). This silence occurs despite – or, more cynically, because of – the pivotal role that institutionalised power relations play in both conditioning how different social groups experience climatic change and in structuring their relative abilities to respond. Without linking localised expressions of vulnerability to broader, historically formed structures of power and privilege, the idea of adaptation can act as a fundamentally depoliticising concept that reduces complex and contested socio-ecological relations to an abstract appeal to defend communities from external environmental disturbances and threats. It is perhaps with this in mind that political ecologists have critiqued mainstream approaches as a "shopping list of 'conditions' for adaptive governance" rather than an analysis of the complex political, cultural and social dynamics at work (Peet, Robbins and Watts 2011a: 9).

Undoubtedly, some critical perspectives within the adaptation literature have sought to challenge technocratic readings of adaptation and focus instead on preexisting social differentiation as a vector of vulnerability to climate change (e.g. O'Brien, St. Clair and Kristoffersen 2010; Pelling 2011). An emerging body of literature that examines vulnerability from a broadly 'human security' perspective, for example, draws on older contributions to the hazards literature to emphasise how the worst affected by abrupt climatic change are disproportionately drawn from segments of society that are chronically marginalised in daily life (see Chapter Four). As Neil Adger notes: "It becomes clear that environmental and social change does not affect everyone equally. Less resilient communities – and more vulnerable individuals – can be severely affected by change, thus limiting their opportunities for adaptation" (Adger and Jordan 2009: 10). This is an important, if somewhat tautological, point. It moves us towards a more progressive politics of adaptation that places due emphasis on how the experience of climate change is shaped by a range of social factors such as the degree of social inequality and political representation. Given the heavily technocratic inclination of much of the adaptation literature, this is an important achievement in and of itself.

Factoring inequality into the framework, however, is not the same as analysing power. The former seeks to understand differential abilities to adapt in terms of a stratified distribution of assets or access to public resources. It takes a given and unequal state of affairs and plots their impact in terms of differential levels of vulnerability or adaptive capacity. A focus on power, in contrast, seeks to understand the dynamic relationships and processes through which humans and their environments are unequally produced over time. It is to examine how different forms of power are produced and operate across spatial scales that facilitate some actors to influence, profit from and find security, while others are disempowered, marginalised and made vulnerable within the context of ongoing socio-environmental transformations (Blaikie et al. 1994; Bohle, Downing and Watts 1994; Oliver-Smith 2004). From such a perspective, the study of vulnerability cannot be reduced to identifying and categorising 'the vulnerable' as do the ever growing number of vulnerability indices that litter the climate change adaptation debate (Hinkel 2011). Instead, it must focus on the exploration of 'vulnerablisation' as a relational process in which vulnerability is produced and reproduced over time between social groups within the active production of their lived environments (cf. Mosse 2007; Collins 2010; Mosse 2010).[1]

To this end, the field of political ecology has consistently sought to examine the power relations involved in both representing and managing the ecological foundations of contemporary landscapes and livelihoods (Blaikie and Brookfield 1987; Peet and Watts 1996; Watts 2000; Robbins 2012). These tasks are undertaken with the normative goal of denaturalising existing socio-environmental orders to better grasp the uneven distribution of gains and risks arising from deeply fused social and ecological processes. Through such means, political ecology speaks directly and vitally to the key question of who has 'power to adapt', how such power is formed and maintained and at whose potential expense it operates. In seeking to understand the differentiated impacts of climatic change, a political ecology perspective requires us to be acutely sensitive to the multiscalar power dynamics that construct our lived environments and that actively yet unevenly reshape their social and physical landscapes (Mustafa 2005). In particular, it demands that we consider the thorny relational issue of how the insecurity of some might be intimately connected to the relative security of others. As Piers Blaikie and Harold Brookfield once tersely argued, the failure to engage such questions is to overlook how "one person's degradation is another's accumulation" (1987: 14).

Deconstructing 'adaptation' through a relational political ecology of power, therefore, rejects understanding climate change impacts through a set of formal and static categories that apply equally across contexts and scales. This is in stark opposition to the formalism of adaptation analysis, wherein the uniform concepts of vulnerability and adaptive capacity are seen to apply universally from the household to the nation and across social space from agrarian India to metropolitan Indianapolis (see Chapter Three). Instead of such abstract formalism, relational political ecology emphasises that what the adaptation perspective terms 'adaptive capacity' or 'vulnerability' are not intrinsic properties of the subject at hand. Rather, they are instead an expression of complex socio-ecological relations between social groups, classes and genders in which such social agents actively yet unequally seek to transform their lived environments in a given historical context.

To make visible these concerns requires a suitably historical methodology that can chart how contemporary experiences of climatic change overlap, accelerate or interrupt ongoing transformative processes. In the specific context of changing rural livelihoods that provides the central focus of the present book, I position climate change as one further element of dynamic agrarian environments in which the foundations of rural life are continually produced, contested and reshaped by active social and biophysical forces operating across geographic scales (Agrawal and Sivaramakrishnan 2000; Mosse 2003; Mustafa 2005). The latter include the diverse and conflicting agencies that reshape rural landscapes, including the commercialisation of agriculture, changing property relations, forms of capital accumulation, the dynamics of national and regional state formation, macro-projects of environmental engineering, migratory flows, technological change and the emergence of new rural subjectivities and political movements (Bernstein 2010; Hall, Hirsch and Li 2011; Peluso and Lund 2011; Rigg, Salamanca and Parnwell 2012; McMichael 2013; van der Ploeg 2013). It is only within this context of ongoing and dynamic agrarian transformations that we can begin to appreciate the past, present and future impacts of climatic change and what political projects' varied narratives of 'adaptation' may sustain.

Adaptation and the politics of representation

This focus on historically grounded processes of environmental production provides a set of entry points through which to critically assess the scaled power relationships through which climate change is produced and then impacts upon agrarian environments. But where does this leave the concept of adaptation? A acknowledgment is necessary here. The original intention of this book was to provide a grounded political ecology of climate change adaptation that could seek to radicalise the idea of adaptation by placing questions of power at the forefront of analysis. The purpose of such an intervention is to seek to make adaptation more attuned to the needs of the poor and marginalised who are faced with the double burden of existing inequalities coupled to greater risks from climatic change (Eriksen and O'Brien 2007; St. Clair 2010; Pelling 2011; see also, Brown 2011). Between inception and completion, however, it became increasingly clear that the problems encountered in writing questions of power into the adaptation paradigm did not stem simply from an overly narrow a framing of adaptation.

Instead, they appeared to be inherent to the concept itself. Despite its seeming self-evidence in a world of climate change, the concept of adaptation seemed peculiarly resistant to being inscribed with questions of power.

In this respect, the second contribution of political ecology is to help us think critically about how the framework of climate change adaptation produces the issue that it subsequently seeks to resolve. By 'produce', I am not suggesting that anthropogenic climatic change is a fiction invented by wayward scientists, academics and policymakers. On the contrary, the following chapters emphasise that there are expressly important transformations occurring in hydro-climatic processes across varied spatial scales that have challenging implications for the present and future of humanity and the rural poor in particular. Instead, the objective is to challenge the perceived naturalness of 'adaptation' as a concept. To do so necessitates examining how the paradigm of climate change adaptation functions as a discourse; that is, as a set of relations between forms of knowledge, structures of power and institutional practices that together produce specific ways of thinking about and acting upon processes of social and ecological change. The purpose of this task is to ask how the framework of climate change adaptation makes the world legible in ways that both naturalise particular types of social relationships and legitimise definite forms of governance and rule.

Given the seeming self-evidence of adaptation as a response to climate change, this may seem a peculiar assertion. Does not climate change necessitate a comprehensive process of adaptation? Like all conceptual devices, however, the idea of adaptation is not a neutral concept but a social construct that establishes a set of discursive parameters for thinking about climate and society that betray historically grounded intellectual traditions (Williams 1980). From its origins in evolutionary biology, the idea of adaptation has migrated outwards to become a "common-sense default assumption" that is now firmly engrained as part of Western folk wisdom about the world (Harrison 1993: 15). On these travels, it has carried with it an encompassing analytical and political baggage that stems from its foundations in Cartesian rationalism. While such assumptions are buried deep within its analytical foundations, the framework of adaptation is shaped by these legacies and they silently permeate how it represents the issues of climatic change and social transformation. It is only by unpacking these foundational assumptions that we can rethink the ways in which it constrains our understandings of the issues at hand.

In this respect, at its core, the concept of climate change adaptation is founded on a rigid separation between climate and society that mirrors the ontological distinction between the social and natural worlds typical of modernist thought (Castree 2001). Climate and society are represented as two distinct systems or domains – one biophysical and natural, the other cultural and social – that relate to each other through a series of ongoing, reciprocal influences. Having separated the two out on ontological grounds, climate and society are then seen to interact with each other as external entities. Society is seen to influence climate through the release of greenhouse gases that alter atmospheric processes resulting in anthropogenic climate change or what climate scientists refer to as an 'external forcing'. Climate, in turn, impacts back upon society through a series of stimuli, shocks and stresses that range from extreme weather events to more subtle shifts in temperature and precipitation. This ontological division between climate and society provides the discursive grounds upon which adaptation emerges as the means to reconcile the strained relationship between the two. The a priori separation between climate and society calls forth the idea of adaptation by enshrining it as the process of adjustment between two externally related systems that have moved out of synch.

Notwithstanding its ubiquity across the climate change literature, we should be expressly attentive to the political implications of how this framework demarcates its conceptual boundaries. The discursive separation of climate and society leads inexorably towards the representation of climate change as an exogenous force that manifests itself in the form of external shocks to an otherwise independent society. This conceptual framework, I believe, is both limited and limiting. It is perhaps ironic that the discourse of adaptation is founded on intrinsically dualistic foundations just as the notion of 'anthropogenic climate change' appears to challenge them. Consider, for example, the presently salient idea of the 'anthropocene' noted above that seeks to capture an "increasingly anxious awareness of the deep connections between human lives, technologies and the energy flows that link oceans, climate and ecosystems" (MacKenzie 2013: 3). Although the term remains disputed, it nonetheless speaks to the ways in which humans are fundamentally embedded in the production of climate. It highlights the degree to which the supposedly 'natural' category of climate is intrinsically the product of conjoined human and non-human forces that cannot be simply extracted and analysed as mutual influences (Sayre 2012; Head 2014). Under such conditions, the foundational idea of distinct yet interacting natural and social systems seems curiously unsuited to a world in which what we term 'nature' has become increasingly and inextricably produced through human activities and their peculiar social dynamics (Smith 1984; Head and Gibson 2012; Castree 2014).

The presence of such intractably conjoined human and non-human processes makes the exercise of drawing boundaries between the assumed 'natural' and the 'social' worlds deeply problematic. Such troubled conceptual frontiers raise a series of important questions about the frameworks through which we seek to represent the world. Our incipient tendency to gravitate towards the nature–society dichotomy occurs because it helps ground our analyses and simplify the empirical terrains that we struggle to convey. It is of great convenience to be able to label one set of processes as 'natural', another set 'social', and then to accord each with distinct dynamics and plot their subsequent interactions. A world framed according to interacting social and ecological systems with distinct 'internal' dynamics and 'external' influences is a much cleaner place to describe than one of messily bundled and contested assemblages of human and non-human forces operating across spatial scales. Even within the field of political ecology, despite its clearly stated purpose as fusing the study of power relations with ecological processes, there are noted concerns that analysts subsume ecology under social dynamics in an explicitly anthropocentric fashion (see Walker 2005). Indeed, sympathetic scholars have questioned whether political ecology itself risks becoming a form of 'politics without ecology' that narrowly focuses upon the inequitable distribution of natural resources, the power relations that govern their utilisation and the institutional conflicts that they unleash (Zimmerer and Bassett 2003).

The fear is that such conceptual simplifications are poorly suited to describing the complexities of the world around us. As Jason Moore notes, once we dig into the processes through which humans and their environments are produced, the Cartesian view of an inherent nature–society dichotomy becomes both theoretically arbitrary and empirically misleading:

> Try drawing a line around the "social" and the "natural" in the cultivation and consumption of food. In a rice paddy or a wheat field, in a cattle feedlot or on our dinner table, where does the natural process end, and the social process begin? The question itself speaks to the tenuous purchase of our Cartesian vocabulary on the everyday realities that we live, and seek to analyze.
> (Moore 2013: 9)

Moore's example is illustrative of a key point that has become a watchtower at the frontier of political ecology. Given the inherent difficulties faced by Cartesian frameworks, political ecologists have become increasingly attuned to understanding how nature–society dichotomies are not reflections of a static ontological division but instead are discursively produced frameworks for viewing and acting upon the world (Dove 1998; Escobar 1999; Biersack 2006; Braun 2011; Peluso and Vandergeest 2011). Representing nature, therefore, is an intrinsically political process. Differing conceptualisations of nature have been constructed, contested and persistently reconfigured in the service of particular political projects and normative visions of social and environmental transformation (Gregory 2001; Castree 2014). As Raymond Williams once wryly noted, our ideas of nature contain an extraordinary amount of human history (1980: 67). It is, therefore, of little surprise that, although it is rarely made explicit, this ongoing contestation over boundaries between nature and society is very much present within debates about climate change. At an institutional level, the spectre of climate change as an 'out of control' element of nature has served to depoliticise how humans are actively involved in producing the forces to which we are then seemingly beholden to adapt (Swyngedouw 2010).

Society, nature and political ecology

In attempting to deconstruct these embedded ontological categories, political ecology creates a formidable challenge for itself. Tim Bryant captures this point astutely when he notes "the difficulty for political ecology, as with other environmental fields of study, is to specify a nature rendered ever more 'slippery' in an increasingly humanized world" (2001: 167). It is one thing to maintain that the ideas of 'society' and 'nature' are discursive constructions. It is quite another to proceed with concrete analysis once such familiar conceptual foundations have been eroded. Within the broad field of agrarian political economy, for example, causality within processes of agrarian change is almost exclusively viewed as anthropocentric wherein the social categories of class and commodities cleave through nature like a hot knife through butter. To the extent that 'ecology' intervenes, it is generally in terms of an external limiting factor that typically fails to capture the extent to which 'social' forces depend upon the particular material properties and dynamic processes of the non-human world (Moore 2013). There is little room in such explicitly Cartesian frameworks, as Timothy Mitchell notes, to examine "how so-called human agency draws its force by attempting to divert or attach itself to other kinds of energy or logic" (2002: 29). For all its many strengths, the tradition of agrarian political economy tends to represent social processes as invariably writing themselves outwards onto the inert substrate of the non-human world.

To go beyond such unbridled anthropocentrism, we are forced to reassess the notion that humans inhabit a social world of their own that exists parallel to the natural world. Tim Ingold puts this alternate starting point well when he notes that nature "is not a surface of materiality upon which human history is inscribed; rather history is the process wherein both people and their environments are continually bringing each other into being" (2000: 87). This foundation is markedly different from the adaptation framework. Instead of starting with the idea of climate and society as pre-constituted systems or domains that mutually influence each other through external shocks and stimuli, the perspective instead asks how varied assemblages of human and non-human forces are worked together in ways that actively produce environments in both their social and climatic dimensions. The emphasis on production is deliberate because it explicitly seeks to break down the reified categorical separations made between humans and their environments or, in the present context, between society and climate. It emphasises that our world is one of constant, active transformation. Humans and their environments do not simply exist. They are continually brought into being through dynamic transformative processes that are indivisibly 'social' and 'natural'. In this respect, our bodies and the landscapes we inhabit are never finished or complete. They are constantly produced through a field of relationships that, as Eric Swyngedouw puts it, embody interlaced chemical, physical, social, economic, political and cultural processes that are combined in "highly contradictory but inseparable manners" (1999: 446).

To adopt such a perspective is to think reflexively about the complex ways in which human and non-human forces are wrought together in active processes of socio-environmental production. It places attention not on a series of external relationships between society and its natural environment but focuses instead upon the inseparably social and biophysical relations through which lived environments – including their human inhabitants – are brought into being and actively reshaped. The concept of metabolism helps capture this idea of active and continuous socio-ecological production. It highlights the crucial flows, exchanges and transformations of material and energy that are inherent to the ongoing creation of the material world (Fischer-Kowalski 1998; Fischer-Kowalski and Huettler 1999; Swyngedouw 2006). It keenly emphasises how humans and their environments exist in a perpetual process of creation and transformation that is intrinsically and simultaneously biophysical and social. From urban conglomerations to rural fields, the physical forms of the environment crystallise as moments of this continual socio-ecological production. In taking such metabolisms seriously, we must be keenly attentive to the ways that our environments are actively constituted through socio-ecological processes operating on different spatial scales and along distinct temporal horizons. As Don Mitchell terms it, scale provides a means to see "how the violent destruction of landscape (and livelihood) in one place can redound very much to the benefit of landscapes (and people) in other places" (2003: 791).[2]

To emphasise non-human forces as productive agents is not an attempt to flatten out the world by assigning equal agency to all manner of human and non-human actors in some form of radical pluralism. Humans are uniquely prolific agents of environmental transformation who, in the process of actively making themselves, continually reshape their lived environments in deeply uneven ways. Within the historical context of capitalism, humans are subjected to a constant drive to churn the earth in search of new materials as part of a series of abstract pressures to continually expand the scale and scope of accumulation (Marx 1973: 164; Harvey 1996; Moore 2010c). This compulsive drive has underscored the rapacious transformation of lived environments on an unparalleled scale and according to increasingly rapid timeframes in which the increasingly intensive commodification of the human and non-human world has been an essential driving force (Smith 1984; Moore 2010b, 2010a). In what we often term the neoliberal era, the extension of institutional frameworks and coercive forces that underlie this expansive commodification has undoubtedly accelerated the pace of environmental production (Taylor 2009). As Jason Moore poses it, "what is finance capital today but a symbolic accounting and material practice of reshaping global natures in a way favourable to the endless accumulation of capital?" (2013: 6). The identification and analysis of such processes, therefore, is central to understanding contemporary environmental flux, including the way in which global climate change is driven by the 'overproduction' of material by-products within an unceasingly expansive industrial metabolism geared towards the relentless accumulation of capital.

Yet we also need to be modestly cautious lest we overstate the power of the anthropogenic. The ability of humans to act, to accumulate and to transform the world around them depends on a field of relations that inherently involves non-human agencies that often betray an unruly resistance to anthropogenic intentions (Latour 1993; Kaika 2005; Bennett 2010). While the spectre of climate change is indeed testament to the material drive of capitalism and the transformative power of human labour, it simultaneously signals tensions inherent to the complex metabolic relationships between humans and the non-human world. The enormous social powers stamped with the imprint of capital appear ever more uneasily hinged upon flows and transformations of energy and matter that remain only partially harnessed to such terse social logics (Mitchell 2002,

2011; Moore 2013). This is to put a further inflection on Marx's notion that, in its propensity to conjure up colossal forces of production, capitalist society is like the sorcerer unable to control the perilous powers raised by his spells (Marx and Engels 1998). If nothing else, the overleveraged geophysical energies visible within contemporary climate change are due warning of the nightmare hubris stemming from humanity's overproduction of the non-human world.

The production of lived environments

My primary purpose in this text, however, is not to examine the macro-contradictions of industrial capitalism per se but to focus more immediately on regional processes of socio-ecological production and climatic change within agrarian Asia. At this level of abstraction, to foreground processes of active environmental production is to break with the idea of the environment as something that is external to humans and which exists as either a material tapestry upon which they write their history or as an external system that provides objective constraints to such agency. Far from occupying a situation of externality, humans are inherently embedded as one element within a field of relations through which both they and their environments are simultaneously produced. Humans do not, therefore, act *upon* their environments. They act *within* them as part of a dense network of interrelationships that are mutually transformative. Ingold puts this eloquently when he remarks that what we have been accustomed to calling "the environment" is better envisaged as a "domain of entanglement" in which human and non-human forces are inseparably bound up in the processes of producing each other. Environments are constantly and perpetually in transition, and humans reproduce themselves as part of these processes. On such terms, the agency of humans is not produced through their separation from their environment. It emerges precisely through their fundamental embeddedness within its active field of relations.

To capture how humans and their environments are mutually constitutive in an active and ongoing manner, I deploy the notion of 'lived environments'. This term extends the concept of agrarian environments levied by Arun Agrawal and K. Sivaramakrishnan (2000). In rejecting both the idea of an autonomous nature that stands outside of society and that of a self-constituting human agency that imposes itself upon such external nature, these authors sought to push forward the frontiers of political ecology. To do so, they highlighted how the physical characteristics, social relations, and cultural representations of agrarian landscapes exist as part of an interlinked process of construction and transformation over time. On this basis, agrarian environments are conceptualised as a dense field of socio-ecological relations operating across spatial scales. They are situated as "part of a biophysical and social environment that always includes the urban and the nonurban, the arable and the non-arable, and other areas that are integrally linked to the world of agriculture and environment and their allied social-economic relations" (Agrawal and Sivaramakrishnan 2000: 8). Following in their tracks, I deploy the notion of lived environments not to refer to a particular physical landscape or locality but rather to emphasise the social and biophysical field of relationships that actively bring such landscapes into being. As Ingold puts it, "the forms of the landscape . . . emerge as condensations or crystallizations of activity within a relational field" (2011: 47). It is this relational field that the idea of lived environments seeks to convey.

To focus on lived environments, therefore, is to emphasise the multi-scalar networks and relationships through which materials, energy, bodies, commodities, capital, pollutants and knowledge are circulated and combined in order to facilitate socio-ecological production (Swyngedouw 2006). As an example, consider the fields of rice plants (paddy) that are common to many agrarian settings across South Asia. Here we encounter a process of seed germination, development, ripening and harvesting that takes form through an assemblage of dynamic forces including the meteorological phenomena through which sun and shade, water and heat are manifested; the human reshaping of water flows to channel and flood specific areas of land; the social patterns of seeding and weeding and their associated divisions of labour and embedded knowledge; the historical engineering of seeds through genetic manipulation via selective breeding or biotechnology; the frequent additions of natural or synthetic chemicals that shift the nutrient and biotic balance of the paddy; the networks of credit provision, land tenure and labouring bodies that shape agricultural production; and the circuits of capital accumulation, market shifts and governmental policies within which rice production is integrally assimilated. From this brief example, we can begin to grasp the complexity of this relational field within which the varied elements of a lived environment are mutually involved in bringing the others into being. It is not simply grains of rice that are actively produced through such socio-ecological processes. On the contrary, we witness the simultaneous production of landscapes, plants, microbes, insects, methane, fodder, livelihoods, profits, forms of knowledge, capital and institutionalised structures of power.

Arising from such diverse and contrasting agencies and intentions, lived environments are inherently fluid and contested. They uneasily combine diverse temporalities and logics, both human and non-human. Rice production, for example, must seek to reconcile varied temporalities ranging from the ever-quickening pulses of capital accumulation, the inherent seasonality of agricultural cycles, the fluctuating needs of household cash flows, to the cyclical surges of insect pests that obdurately learn to resist the varied human attempts to control them. Equally they synthesise agencies working at different scales and with vastly different concentrations of power: from the level of household livelihood through to the decisions of bureaucrats at the World Trade Organisation and executives in corporate biotech labs. It is perhaps no surprise, therefore, that the institutional scaffoldings that humans construct in an attempt to gird lived environments frequently strain and buckle under the weight of their embedded contradictions and the multiple struggles to which they give rise. As the historian David Ludden has pointedly noted, such environments emerge as geographical spaces defined by social power and resistance that "together produce and transform entitlements such as the rights to use land, water, forests and other collective property" (2002: 239). A resolutely historical perspective, therefore, emerges as an essential means to understand how the contending forces through which lived environments are continually co-constructed, contested and remade.

The climate of lived environments

To approach the issue of climatic change from the perspective of lived environments impels us to rethink the category of climate that is at the heart of the issue. As noted above, within the adaptation paradigm, climate is commonly cast as an external system that impacts upon humans and their environments through localised shocks, stresses and stimuli. This representation builds upon a rationalist framing of climate as something uniquely biophysical that exists in separation from society as an influencing factor and a limiting constraint (see Chapter Two). By taking climate and society as given external elements, however, the adaptation paradigm obscures precisely what is most important. It fails to interrogate the ways that both climate and society do not simply exist in an ontological sense but are actively and continually brought into being through processes that are indivisibly social and natural, discursive and material. As I demonstrate both conceptually and through case material from rural Asia, climates take shape through the embedding of meteorological forces within a wider set of socio-ecological processes. Meteorological phenomena such as precipitation, wind, temperature and light are unevenly written into socio-ecological orderings of land, bodies, plants, capital, infrastructure, technologies and knowledge in ways that produce crops, fuel, fibre and other materials essential for social reproduction. It is how such meteorological forces are worked into this wider field of socio-ecological relations in any given location that makes climates real and tangible. In short, climates in this substantive material sense are – *in part* – socially produced.

This approach requires us to consider historically how climates are produced through the working of meteorological forces into the production of a lived environment, including its physical landscapes, its built infrastructures and its social hierarchies. The manifestation of precipitation in an urban centre in interior India, for example, is qualitatively distinct from its rural hinterland despite the two being separated by mere kilometres. The very same quantitative amount of rain that registers on the weather statistics derived from both locations obscures how rainfall takes on different purposes, drives different socio-ecological processes and has very different cultural meanings in each environment. In the urban area, the rainy season may be experienced as a simple relief from the summer heat or as a watery threat to habitation and livelihood, depending on the specific construction of the urban form and the segregation of marginal bodies within it. In the rural area, rain – or its absence – is a life-or-death question that becomes tangible in relation to the specific couplings of crops, labouring patterns, forms of infrastructure, land tenure arrangements, disbursements of credit and potential access to other livelihoods. In particular, monsoon storms bring fluidity not just to the drainage channels that irrigate the fields but also to the social relations of credit and debt through which surpluses and risks are constructed and distributed. In short, the very same abstract climatic trends manifest themselves in radically different ways in these two locations because they are materialised in and through a specific relational field. The two places have very different material climates.

If we therefore conceive that climate is – in part – socially produced through the way that meteorological processes form part of wider socio-ecological assemblages, then we need to rethink the idea of climate change adaptation. The focus on such socio-ecological dynamics gravely complicates the adaptation framework because it undermines the idea of clearly defined boundaries between climate and society upon which the idea of adaptation is drawn. Instead of a clear process of social adaptation to an external climate system, we are forced to grapple with the complex couplings of human and non-human forces through which lived environments in both their social and climatic dimensions are perpetually formed and transformed. It is only by understanding how meteorological forces are situated within a historically specific field of socio-ecological production that we can grasp their tangible role as active and productive elements of a lived environment. The inverse side of such productive forces, however, is that the very same couplings of socio-ecological processes with meteorological forces simultaneously create dynamic and strikingly uneven landscapes in which such combinations can act as conduits of considerable destruction. The boundary line from life-giving rains to destructive floods is often a fine one, hinged upon the specific ways in which meteorological and social forces are brought together within the production of a lived environment. This requires a suitably historical approach that can situate meteorological forces within broader socio-ecological processes and their underlying power relations. I develop this argument with specific respect to glacial outbursts in Uttarkhand (Chapter Three), floods in Pakistan (Chapter Six), drought in India (Chapter Seven) and winter storms in Mongolia (Chapter Eight).

Political ecology beyond adaptation

The above argument forms the basis for the critique of climate change adaptation as an analytical framework and a foundation for political action. If, as the book argues, climate is not something 'out there' but is actively produced as an essential moment of the formation of lived environments, the discursive boundaries upon which the adaptation paradigm rests become tenuous. What is termed anthropogenic climatic change is no more 'natural' or 'external' than the appearance of pesticide-resistant insects in an agrarian environment or the coat of smog that blankets innumerable cities from Baltimore to Beijing. Each is an outcome of complex forms of socio-ecological production that operate across varied spatial scales, temporal horizons and social divides. From this perspective, our attention becomes focused not on an 'out of control' global climate that exists as a coherent external power to which we need to adapt. Instead, we must ask how our lived environments, in both their social and climatic dimensions, are actively produced through the complex interaction of human and non-human agencies in ways that are markedly unequal. This leads towards a fundamental political shift. Engaging contemporary climatic change is not about adapting to a changing external environment. It is about challenging how we produce ourselves. Instead of a politics of adaptation, we need a politics of producing ourselves differently.

The adaptation framework, however, is intrinsically resistant to exploring such complex social-ecological relations and their political implications. Through the representation of climate as an external influence upon an otherwise coherent society, it is geared towards the discursive production of climatic change as a bounded, external and ultimately governable phenomenon. This representational strategy not only creates relatively standardised conventions for talking about climate change, it simultaneously produces legitimating frameworks upon which managerial interventions and technocratic governance can be facilitated and rationalised. In exploring these tensions in the following chapters, I argue that climate change adaptation does not present a neutral conceptual framework that can simply be filled with a more progressive content as needed. It is a discourse that is intrinsically embedded in, and reproductive of, material forces, institutionalised practices and political claims that are closely geared towards the preservation of existing social and environmental parameters. Such constraints, for example, are acutely manifested in the mainstream discourse's deeply problematic separation of 'adaptation' and 'mitigation', which proceeds from an assumption that the experience of climatic change can be adequately envisaged and engaged in separation from the processes that produce it.

Climate change, however, is not something that can be separated out, managed and governed as an external influence upon a pre-existing society. It exists as one further moment within the scaled processes through which our lived environments, with their vast inequalities and engrained forms of power, are actively produced, contested and transformed. In breaking with the adaptation approach, we require a suitably broad perspective that can situate climatic change in long-standing historical processes of social and environmental transformation that are inextricably written into the dynamic socio-ecology of contemporary capitalism. It is in this respect that the framework of political ecology helps us connect the issue of climate change to the broader social struggles that animate our lived environments. Although obscured within the adaptation literature, conflicts over resources and livelihoods in conditions of sharp environmental shifts are deeply engrained within the social fabric of both urban and agrarian life across much of the global South. Within agrarian south Asia, for example, the latter have been recognised drivers of social mobilisation and historical change from pre-colonial times through to the present (Gadgil and Guha 1993; Baviskar 2001; Davis 2002; Mosse 2003; Rajan 2006). Bringing these historical power dynamics and socio-political processes into discussion of climate change are, I believe, essential for a closer conceptualisation of the forces and relationships that drive social and ecological change and unevenly distribute both its risks and rewards. Ultimately, they offer quite a different basis upon which to begin re-envisioning how we might differently produce our lived environments within the context of climatic change.

Notes

1 It is conspicuous how many present writings on climate change adaptation duly reference the first generation of political ecology literature while remaining impervious to its primary arguments. The seminal article by Watts and Bohle (1993) on vulnerability and entitlements in the context of climatic change, for example, is widely cited across the adaptation literature while its central argument – that a theory of vulnerability needs not only to examine individual command over resources and basic necessities but also the structural properties of the political economy as a whole – is assiduously ignored.

2 In this respect, early political ecology accounts commonly focused on explaining a relatively localised set of dynamics surrounding, for example, land degradation or irrigation management, by locating them within political economic processes working at different scales (Blaikie 1985; Blaikie and Brookfield 1987). While duly highlighting the importance of scale, this approach tended towards a static and mono-directional framework in which the study of environmental change was situated within a preconstituted scalar hierarchy layered from global to local with privilege often accorded to the causative role of the former (Watts 2000). Frustration with the limitations of such a perspective underpinned justified concerns regarding the structuralist tendencies of the earlier contributions (see Blaikie 1997 for a self-critique). Subsequent work has become increasingly focused not on scale as a natural, pre-given dimension of human practice within which social and ecological processes unfold but rather on the construction and contestation of scale through specific socio-ecological relations and networks (Smith 1984; Swyngedouw 1997; MacKinnon 2010; Birkenholtz 2011).

References

Adger, W. N., and A. Jordan. 2009. *Governing Sustainability*. Cambridge, New York: Cambridge University Press.

Adger, W. N., I. Lorenzoni, and K. O'Brien. 2010. "Adaptation Now." Pp. 1–22 in *Adapting to Climate Change*, edited by W. Neil Adger, I. Lorenzoni, and K. O'Brien. Cambridge: Cambridge University Press.

Agrawal, A., and K. Sivaramakrishnan. 2000. "Introduction: Agrarian Environments." Pp. 1–22 in *Agrarian Environments: Resources, Representation and Rule in India*, edited by Arun Agrawal and K. Sivaramakrishnan. Durham, NC: Duke University Press.

Barbier, E. 2011. *Scarcity and Frontiers*. Cambridge: Cambridge University Press.

Bassett, T., and C. Fogelman. 2013. "Deja vu or something new? The adaptation concept in the climate change literature." *Geoforum* 43(1): 42–54.

Baviskar, A. 2001. "Written on the Body, Written on the Land: Violence and Environmental Struggles in Central India." Pp. 354–79 in *Violent Environments*, edited by Nancy Peluso and Michael Watts. Ithaca, NY: Cornell University Press.

Bennett, J. 2010. *Vibrant Matter: A Political Ecology of Things*. Durham, NC: Duke University Press.

Bernstein, H. 2010. *Class Dynamics of Agrarian Change*. Halifax, NS: Fernwood.

Biersack, A. 2006. "Reimagining Political Ecology: Culture/Power/History/Nature." Pp. 3–40 in *Reimagining Political Ecology*, edited by Aletta Biersack and James Greenberg. Durham, NC: Duke University Press.

Birkenholtz, T. 2011. "Network political ecology: Method and theory in climate change vulnerability and adaptation research." *Progress in Human Geography* 36(3): 295–315.

Blaikie, P. M. 1985. *The Political Economy of Soil Erosion in Developing Countries*. London: Longman.

Blaikie, P. M. 1997. "Classics in human geography revisited." *Progress in Human Geography* 21(1): 75–80.

Blaikie, P. M. 2001. "Social Nature and Environmental Policy in the South: Views from Verandah and Veld." Pp. 133–50 in *Social Nature*, edited by Noel Castree and Bruce Braun. Oxford: Blackwell.

Blaikie, P. M., and H. C. Brookfield. 1987. *Land Degradation and Society*. London: Methuen.

Blaikie, P. M., J. Cameron, and D. Seddon. 1983. *Centre, Periphery, and Access in West Central Nepal: Approaches to Social and Spatial Relations of Inequality*. Norwich, UK: School of Development Studies, University of East Anglia.

Blaikie, P. M., T. Cannon, I. Davis, and B. Wisner. 1994. *At Risk: Natural Hazards, People's Vulnerability, and Disasters*. London: Routledge.

Bohle, H., T. Downing, and M. Watts. 1994. "Climate change and social vulnerability." *Global Environmental Change* 4(1): 37–48.

Braun, B. 2011. "Governing Disorder: Biopolitics and the Molecularization of Life." Pp. 389–411 in *Global Political Ecology*, edited by Richard Peet, Paul Robbins, and Michael Watts. London: Routledge.

Bryant, T. 2001. "Political Ecology: A Critical Agenda for Change?" Pp. 151–69 in *Social Nature: Theory, Practice and Politics*, edited by Noel Castree and Bruce Braun. Oxford: Blackwell.

Brown, K. 2011. "Sustainable adaptation: An oxymoron?" *Climate and Development* 3(1): 21–31.

Castree, N. 2001. "Socializing Nature: Theory, Practice and Politics." Pp. 1–21 in *Social Nature: Theory, Practice and Politics*, edited by Noel Castree and Bruce Braun. Oxford: Blackwell.

Castree, N. 2014. *Making Sense of Nature: Representation, Politics and Democracy*. London: Routledge.

Chakrabarty, D. 2009. "The climate of history: Four theses." *Critical Inquiry* 35(4): 197–226.

Clark, N. 2010. "Volatile worlds, vulnerable bodies: Confronting abrupt climate change." *Theory, Culture and Society* 27(2–3): 31–53.

Collins, T. 2010. "Marginalization, facilitation and the production of unequal risk: The 2006 Paso del Norte Floods." *Antipode* 42(2): 258–88.

Crutzen, P., and W. Steffen. 2003. "How long have we been in the Anthropocene era?" *Climate Change* 61(1): 251–57.

Davis, M. 2002. *Late Victorian Holocausts*. London: Verso.

Dove, M. 1998. "Living rubber, dead land, and persisting systems in Borneo: Indigenous representations of sustainability." *Journal of the Humanities and Social Sciences of Southeast Asia and Oceania* 154(1): 20–54.

Eriksen, S., and K. O'Brien. 2007. "Vulnerability, poverty and the need for sustainable adaptation measures." *Climate Policy* 7(4): 337–52.

Escobar, A. 1995. *Encountering Development: The Making and Unmaking of the Third World*. Princeton, NJ: Princeton University Press.

Escobar, A. 1999. "After nature: Steps to an antiessentialist political ecology." *Current Anthropology* 40(1): 1–30.

Fagan, B. M. 2004. *The Long Summer: How Climate Changed Civilization*. New York: Basic Books.

Fischer-Kowalski, M. 1998. "Society's metabolism: The intellectual history of materials flow analysis, part I, 1860–1970." *Journal of Industrial Ecology* 2(1): 61–79.

Fischer-Kowalski, M., and W. Huettler. 1999. "Society's metabolism: The intellectual history of materials flow analysis, part II, 1970–1998." *Journal of Industrial Ecology* 2(4): 107–37.

Gadgil, M., and R. Guha. 1993. *This Fissured Land: An Ecological History of India*. Berkeley: University of California Press.

Gregory, D. 2001. "(Post)Colonialism and the Production of Nature." Pp. 84–111 in *Social Nature: Theory, Practice, Politics*, edited by Noel Castree and Bruce Braun. Oxford: Blackwell.

Grove, R. 1997. *Ecology, Climate and Empire: Colonialism and Global Environmental History, 1400–1940*. Cambridge: White Horse Press.

Hall, D., P. Hirsch, and T. Li. 2011. *Powers of Exclusion: Land Dilemmas in Southeast Asia*. Honolulu: University of Hawaii Press.

Harrison, G. A. 1993. *Human Adaptation*. Oxford: Oxford University Press.

Harvey, D. 1996. *Justice, Nature and the Geography of Difference*. Oxford: Blackwell.

Head, L. 2010. "Cultural ecology: adaptation – retrofitting a concept?" *Progress in Human Geography* 34(2): 234–42.

Head, L. 2014. "Contingencies of the Anthropocene: Lessons from the 'Neolithic'." *The Anthropocene Review* forthcoming.

Head, L., and C. Gibson. 2012. "Becoming differently modern: Geographic contributions to a generative climate politics." *Progress in Human Geography* 36(6): 699–714.

Hinkel, J. 2011. "Indicators of vulnerability and adaptive capacity: Towards a clarification of the science-policy interface." *Global Environmental Change* 21: 198–208.

Hoselitz, B. 1952. "Non-economic barriers to economic development." *Economic Development and Cultural Change* 1(8): 8–22.

Hulme, M. 2010. "Cosmopolitan climates: Hybridity, foresight and meaning." *Theory, Culture and Society* 27(2–3): 267–76.

IFAD. 2010. *Rural Poverty Report 2011*. Rome: IFAD.

IFAD. 2013. *The Adaptation Advantage: The Economic Benefits of Preparing Small-Scale Farmers for Climate Change*. Rome: IFAD.

Ingold, T. 2000. *The Perception of the Environment: Essays on Livelihood, Dwelling and Skill*. London: Routledge.

Ingold, T. 2011. *Being Alive: Essays on Movement, Knowledge and Description*. London: Routledge.

IPCC. 2007. *Climate Change 2007: The Scientific Basis. WG I Contribution to IPCC 4th Assessment Report*. Cambridge: Cambridge University Press.

Ireland, P. 2012. "Climate change adaptation: Business-as-usual aid and development or an emerging discourse for change?" *International Journal of Development Issues* 11(2): 92–110.

Kaika, M. 2005. *City of Flows: Modernity, Nature and the City*. London: Routledge.

Latour, B. 1993. *We Have Never Been Modern*. Cambridge, MA: Harvard University Press.

Leach, M., R. Mearns, and I. Scoones. 1999. "Environmental entitlements: Dynamics in community-based natural resource management." *World Development* 27(2): 225–47.

Leary, N., J. Adejuwon, V. Barros, P. Batima, B. Biagini, I. Burton, … M. Wehbe. 2010. "A Stitch in Time: General Lessons from Specific Cases." Pp. 1–27 in *Climate Change and Adaptation*, edited by Neil Leary et al. London: Earthscan Press.

Ludden, D. 2002. "Spectres of agrarian territory in southern India." *Indian Economic Social History Review* 39(2&3): 233–59.

MacKenzie, A. 2013. "Having an anthropocene body: Hydrocarbons, biofuels and metabolism." *Body and Society* 20(1): 3–30.

MacKinnon, D. 2010. "Reconstructing scale: Towards a new scalar politics." *Progress in Human Geography* 35(1): 21–36.

Martínez-Alier, J. 2002. *The Environmentalism of the Poor : A Study of Ecological Conflicts and Valuation*. Cheltenham, UK: Edward Elgar.

Marx, K. 1973. *Grundrisse*. London: Penguin Books.

Marx, K., and F. Engels. 1998. *The Communist Manifesto*. New York: Monthly Review Press.

McMichael, P. 2013. *Food Regimes and Agrarian Questions*. Halifax, NS: Fernwood.

Mitchell, D. 2003. "Cultural landscapes: Just landscapes or landscapes of justice?" *Progress in Human Geography* 27(6): 787–98.

Mitchell, T. 2002. *Rule of Experts: Egypt, Techno-Politics, Modernity*. Berkeley: University of California Press.

Mitchell, T. 2011. *Carbon Democracy: Political Power in the Age of Oil*. London: Verso Books.

Moore, J. 2010a. "'Amsterdam is standing on Norway' part I: The alchemy of capital, empire and nature in the diaspora of silver." *Journal of Agrarian Change* 10(1): 33–68.

Moore, J. 2010b. "'Amsterdam is standing on Norway' part II: The global North Atlantic in the ecological revolution of the long seventeenth century." *Journal of Agrarian Change* 10(2): 188–227.

Moore, J. 2010c. "The end of the road? Agricultural revolutions in the capitalist world-ecology, 1450–2010." *Journal of Agrarian Change* 10(3): 389–413.

Moore, J. 2013. "From Object to Oikeios: Environment-Making in the Capitalist World-Ecology." Unpublished paper. Department of Sociology, Binghamton University. Available at: http://www.jasonwmoore.com/uploads/Moore__From_Object_to_Oikeios__for_website__May_2013.pdf

Mosse, D. 2003. *The Rule of Water: Statecraft, Ecology and Collective Action in South India*. Oxford: Oxford University Press.

Mosse, D. 2007. "Power and the durability of poverty: a critical exploration of the links between culture, marginality and chronic poverty." CPRC Working Paper, Anthropology Department, SOAS (107): 1–60.

Mosse, D. 2010. "A relational approach to durable poverty, inequality and power." *Journal of Development Studies* 46(7): 1156–78.

Mustafa, D. 2005. "The production of an urban hazardscape in Pakistan: Modernity, vulnerability, and the range of choice." *Annals of the Association of American Geographers* 95(3): 566–86.

Neumann, R. 2005. *Making Political Ecology*. London: Hodder Arnold Press.

O'Brien, K., A. L. St. Clair, and B. Kristoffersen (Eds.). 2010. *Climate Change, Ethics and Human Security*. Cambridge: Cambridge University Press.

Oliver-Smith, A. 2004. "Theorizing Vulnerability in a Globalized World: A Political Ecological Perspective." Pp. 10–25 in *Mapping Vulnerability: Disasters, Development and People*, edited by Greg Bankoff, Georg Frerks, and Dorothea Hilhorst. London: Earthscan.

Peet, R., P. Robbins, and M. Watts. 2011a. "Global Nature." Pp. 1–48 in *Global Political Ecology*, edited by Richard Peet, Paul Robbins, and Michael Watts. London: Routledge.

Peet, R., P. Robbins, and M. Watts. 2011b. *Global Political Ecology*. Abingdon, Oxon; New York: Routledge.

Peet, R., and M. Watts. 1996. *Liberation Ecologies: Environment, Development, Social Movements*. London; New York: Routledge.

Peet, R., and M. Watts (Eds.). 2004. *Liberation Ecologies: Environment, Development, Social Movements*. London: Routledge.

Pelling, M. 2011. *Adaptation to Climate Change*. London: Routledge Press.

Peluso, N. L., and C. Lund. 2011. "New frontiers of Land Control." *Journal of Peasant Studies* 38(4): 667–81.

Peluso, N. L., and P. Vandergeest. 2011. "Taking the Jungle out of the Forest: Counter-Insurgency and the Making of National Natures." Pp. 252–84 in *Global Political Ecology*, edited by Richard Peet, Paul Robbins, and Michael Watts. London: Routledge.

Rajan, S. R. 2006. *Modernizing Nature: Forestry and Imperial Eco-Development 1800–1950*. Oxford: Oxford University Press.

Rappaport, R. A. 1979. *Ecology, Meaning, and Religion*. Richmond, CA: North Atlantic Books.

Rigg, J., A. Salamanca, and M. Parnwell. 2012. "Joining the dots of agrarian change in Asia: A 25 year view from Thailand." *World Development* 40(7): 1469–81.

Robbins, P. 2000. "Pastoralism and Community in Rajasthan: Interrogating Categories of Arid Lands Development." Pp. 191–215 in *Agrarian Environments: Resources, Representation and Rule in India*, edited by Arun Agrawal and K. Sivaramakrishnan. Durham, NC: Duke University Press.

Robbins, P. 2012. *Political Ecology: A Critical Introduction. (2nd edition)*. Oxford, UK: Blackwell.

Sayre, N. 2012. "The politics of the anthropogenic." *Annual Review of Anthropology* 41(1): 57–70.

Smith, N. 1984. *Uneven Development: Nature, Capital, and the Production of Space*. New York: Blackwell.

St. Clair, A. L. 2010. "Global Poverty and Climate Change: Towards the Responsibility to Protect." Pp. 180–98 in *Climate change, Ethics and Human Security*, edited by Karen O'Brien, Asuncion Lera St. Clair, and Berit Kristoffersen. Cambridge: Cambridge University Press.

Swyngedouw, E. 1997. "Neither Global nor Local: 'Glocalisation' and the Politics of Scale." Pp. 137–66 in *Spaces of Globalization*, edited by K. Cox. New York: Guilford.

Swyngedouw, E. 1999. "Modernity and hybridity: Nature, regeneracionismo, and the production of the Spanish waterscape, 1890–1930." *Annals of the Association of American Geographers* 89(3): 443–65.

Swyngedouw, E. 2004. *Social Power and the Urbanization of Water: Flows of Power*. Oxford; New York: Oxford University Press.

Swyngedouw, E. 2006. "Circulations and metabolisms: (Hybrid) natures and (cyborg) cities." *Science as Culture* 15(2): 105–21.

Swyngedouw, E. 2010. "Apocalypse forever? Post-political populism and the spectre of climate change." *Theory, Culture and Society* 27(2–3): 213–32.

Swyngedouw, E., and N. Heynen. 2003. "Urban political ecology, justice and the politics of scale." *Antipode* 35(5): 898–918.

Taylor, M. 2009. "Displacing insecurity in a divided world: Global security, international development and the endless accumulation of capital." *Third World Quarterly* 30(1): 147–62.

UNDP. 2007. *Human Development Report 2007/2008: Fighting Climate Change – Human Solidarity In A Divided World*. Geneva: United Nations.

UNEP. 2014. *Assessing Global Land Use: Balancing Consumption with Sustainable Supply*. Geneva: United Nations Environment Programme.

van der Ploeg, J. 2013. *Peasants and the Art of Farming: A Chaynovian Manifesto*. Halifax, NS: Fernwood.

Walker, P. 2005. "Political ecology: Where is the ecology?" *Progress in Human Geography* 29(1): 73–82.

Watts, M. 1983. *Silent Violence: Food, Famine and Peasantry in Northern Nigeria*. Berkeley: University of California Press.

Watts, M. 2000. "Political Ecology." Pp. 257–75 in *Companion to Economic Geography*, edited by Eric Sheppard and Trevor Barnes. Oxford: Blackwell.

Watts, M. 2004. "Antinomies of community: Some thoughts on geography, resources and empire." *Transactions of the Institute of British Geographers* 29(2): 195–216.

Watts, M., and H. Bohle. 1993. "The space of vulnerability: the causal structure of hunger and famine." *Progress in Human Geography* 17(1): 43–69.

Williams, R. 1980. *Problems in Materialism and Culture*. London: Verso.

Wolf, E. R. 1982. *Europe and the People without History*. Berkeley: University of California Press.

World Bank. 2008. *Climate Change Impacts in Drought and Flood Affected Areas: Case Studies in India*. Washington, DC: IDRC.

Zimmerer, K., and T. Bassett. 2003. "Approaching Political Ecology: Society, Nature and Scale in Human-Environment Studies." Pp. 1–28 in *Political Ecology: An Integrative Approach to Geography and Environment Development Studies*, edited by Karl Zimmerer and Thomas Bassett. New York: Guildford Press.

2 Socialising climate

A curious facet of the literature on climate change adaptation is its collective silence regarding the idea of climate that stands at its foundation. Notwithstanding the voluminous work debating concepts such as vulnerability, adaptive capacity and resilience, climate is largely conspicuous by its absence. This pervasive ambiguity surrounding a key foundational term gives rise to the impression that climate is perhaps a self-evident category that stands in no need of conceptual elucidation. Such an assumption, I contend, is deeply misleading. How we conceptualise climate greatly shapes the parameters of what we subsequently consider to be climate change, its causative dynamics and the potential implications we draw for social and environmental transformation. As a result, the process of 'fixing' climate, by which we draw conceptual boundaries around various geophysical and social processes so as to delineate climate as a coherent domain, is an inherently political exercise. It involves ordering complex socio-ecological phenomena into distinct and separated categories of 'climate' and 'society', for which the underlying casual determinants might then be represented as 'natural' or 'human'. Such discursive processes greatly shape how we understand 'anthropocentric climate change' in the present, yet the political dimensions of fixing climate are largely undertaken in silence.[1]

In what follows, I argue that the unwillingness of adaptation theorists to explicitly address the underlying idea of climate has led to the tacit incorporation of a rationalist framework sourced directly from the natural sciences. The latter casts climate as something natural and physical – a composite of meteorological indicators – that can be understood in abstraction from the social world that it subsequently impacts upon. It is this very separation of climate and society that gives coherence to the idea of adaptation as a process of social adjustment to a changing external environment. In response, the present chapter reconstructs an alternative framing of climate that emphasises its inherent social dimensions. By rejecting the idea of climate as an external biophysical domain of natural atmospheric processes, I argue that climatic change is better understood from the perspective of 'material climates' which emphasises the interlacing of meteorological forces, forms of social organisation, technological infrastructures and discursive frameworks at various spatial scales. By refusing to pre-package these socio-ecological processes into bounded ontologies of 'climate' and 'society', this approach opens the way for a more nuanced conceptualisation of the forms of power through which climates are produced and that render them both productive – and sometimes deeply destructive – of human life.

Scientific rationalism and the externalisation of climate

In reading the assorted literature on climate change and adaptation, it is necessary to dig deep to uncover an explicit elaboration of the concept of climate. The Intergovernmental Panel on Climate Change (IPCC), for example, buries its definition of climate within the appendix section of its various assessment reports. There it references the World Meteorological Organization with a short definition that states climate is the "average weather" or, more expansively, "the statistical description in terms of the mean and variability of relevant quantities over a period of time ranging from months to thousands or millions of years" (IPCC 2007: 869). The idea of climate, in this framing, represents a register of meteorological processes that can be collectively rendered legible through the measurement and averaging of various physical indicators, such as variations in temperature, humidity, atmospheric pressure, precipitation, wind and atmospheric particle count (Hulme et al. 2008; Edwards 2010). To produce climate in this way, as Mike Hulme describes, requires an elaborate socio-technical circuitry of measurement and circulation that can statistically capture and process long-term weather trends into the master category of climate:

> By standardising such measurements and then by circulating them through centralised bureaucracies, it became possible, first, to quantify (local) weather and subsequently to construct statistical (aggregated) climates . . . Weather is first captured locally and quantified, then transported and aggregated into regional and global indicators. These indicators are abstracted and simulated in models before being delivered back to their starting places (locales) in new predictive and sterilised forms.
>
> (Hulme 2008: 5)

It is through such socio-technical apparatus that the idea of climate as a composite of uniquely physical processes can be given a tangible form. As a statistical artefact, climate appears in the form of an index of physical weather trends over a selected period of time across a delineated spatial region (Edwards 2010). It, therefore, is produced as an abstraction, an enumerated catalogue of meteorological trends that stands outside of any specific social context and is divorced from the entanglements of human actions with meteorological forces that characterise our lived environments.

This extraction of climate from a social context is, of course, indispensable for the purposes of climate science and its emphasis on modelling. Having produced climate by aggregating localised statistical indicators, these data can then be processed to provide an account of meteorological dynamism and fluctuations, including both cyclical changes and secular transformations. Within climate science, the causative forces behind changing localised climate variables are seen to emerge from what is termed the 'global climate system'. Representing climate in terms of a global system is a sophisticated means to capture a highly complex series of interacting processes that occur across atmospheric levels, the earth's surface and the biosphere. This is done so as to give them order as part of a coherent biophysical totality with its own dynamics and emergent properties. The global climate system, for example, transforms over time either through internal shifts or what are termed 'external forcings' that include "volcanic eruptions, solar variations, or human-induced modifications to the planetary radiative balance, for instance via anthropogenic emissions of greenhouse gases and/or land-use changes" (IPCC 2007: 872). Anthropogenic dynamics, therefore, are represented as a vitally important factor, yet they are seen in terms of an outside influence upon an otherwise coherent system. This consolidates the idea of climate as a bounded domain of nature that exists in abstraction from the social world, resulting in a framework in which society and climate relate to each other as external influences. Indeed, it is precisely by modelling climate in terms of a unitary, planetary scale system of biophysical cause- and-effect that climate science produces a binary in which climate and society appear in terms of two separate systems, one natural, the other social. From this position of externality, the two systems are subsequently seen to influence and effect change in each other.

Although this notion of climate is now silently ubiquitous across the climate change literature, its origins are relatively recent. Its consolidation began in the nineteenth century as part of the broader movement of scientific rationalism that strove to simplify the world into nature on one side and human calculation and expertise on the other (Mitchell 2002: 36). Prior representations of climate operated upon qualitatively different premises. They tended to situate climate geographically, as an interaction of physical processes, human experience and social practice. Vladimir Jankovic, for example, notes how early modern naturalists considered climate as a spatial frame of reference used to categorise and evaluate local features of both nature and society. Rather than a long-term statistical average of chosen atmospheric processes, climate was understood both descriptively to refer to a sum total of human experience and natural production within a given latitude and prescriptively to identify the salient features that unified local topography, biological life and society (Jankovic 2010: 203; see also, Fleming and Jankovic 2011). Emphatically holistic in the way they incorporated various elements of human practice within their understanding of climate, such framings operated within an explicitly regional scale that could capture such intersections between meteorological processes and social process. Although they were undoubtedly beset with differing problems of measurement, interpretation and ideology (Fleming 1998; Heymann 2011), they nonetheless retained conceptual room for the intersection of biophysical processes and social dynamics. Significantly, climate appeared in these framings as something intrinsically natural and social, a point to which I shall return below.

The climatology of scientific rationalism, however, had no truck for the incorporation of the social world within its understanding of nature. In contrast, it established itself on the basis of studying nature 'as it really is', prior to and external from subjective interpretation or human activities (Ingold 2011). Nature, it was contended, could be understood as a system of quantified interrelationships open to human observation and measurement from which their discrete causal dynamics could be extracted. It was on this basis that the idea of climate was increasingly represented as a realm of physical processes that maintained a formal integrity external to the social world and its associated cultural worldviews and productive practices (Anderson 2005). To understand climate as natural and pure, the social world needed to be extracted and separated from it. It was on this basis that climate was increasingly portrayed as part of an external realm of the 'natural environment', separate from the world of human energies yet at the same time constructed as often a determinate influence or constraint upon the latter.[2] The idea of the global climate system is the culmination of this trend.

Although presently dominant, the emergence of this uncompromisingly rationalist climatology was gradual and did not represent simply a cumulative march of scientific advancement. The creation and circulation of new measuring instruments and techniques were undoubtedly a necessary part of its consolidation, yet such technical presuppositions emerged as part of a historically situated political economy of scientific endeavour. In this respect, the rationalist aim to produce nature as a pure, knowable and externalised realm was integral to the project of taming and manipulating the former for human purposes that characterised early industrial capitalism and colonial conquest (Williams 1980). New ideas of climate were intertwined with the dramatic socio-ecological transformations of the nineteenth century wrought by industrialisation, agrarian rationalisation, colonial expansion and military planning, all of which played a role in shaping the staggered evolution of modern climatology (Demeritt 2001; Anderson 2005; Golinsky 2008; Jankovic 2010; Carey 2011, 2012; Heymann 2011). Classifying and rationalising the climate of colonial India, for example, emerged as one of a spectrum of British governmental techniques that incorporated detailed surveying, mapping, cataloguing and census taking (Grove 1997; Prakash 1999; Hazareesingh 2012). Together these forms of knowledge production proved integral to the productive re-ordering of colonial spaces that underscored colonial rule and facilitated the continual flows of products and resources that fed European industrialisation (Gidwani 2008).

In the context of vast social and ecological transformations, governmental requirements for particular types of knowledge helped guide the growth of dynamic meteorology as a separate field that attempted to synthesise newly available weather data into coherent causal models. Climate, as Fleming and Jankovic put it, increasingly began to act as a discursive vehicle capable of turning social questions into "matters of natural fact" so as to rationalise their management (2011: 10). With scientific rationality demanding a clear separation between the realm of social experience and values, on the one hand, and the operation of physical processes with strict laws of causality, on the other, the dynamic meteorology that coalesced by the end of the nineteenth century was an unabashedly reductionist physical science. It was predicated upon a mathematical description of meteorological parameters that shifted the idea of climate further away from the realm of human experience and specific socio-ecological settings into the realm of biophysical models girded firmly to the laws of physics (Heymann 2011). While in the early decades of the twentieth century tensions still existed between an empiricist climatology based on the primacy of interpreting weather data and a theoretical meteorology concerned with the formulation of general laws of causality, these divisions began to recede and ultimately vanished with the rise of computer modelling in the post–World War II period (Edwards 2010).

In shedding the last vestiges of earlier holistic and regionalist approaches, this new climatology consolidated during the twentieth century into what we now term climate science with its emphasis on rationalist causal models that are global in scope. Predicated upon the increasingly powerful mapping of global atmospheric trends as part of an integrated and cohesive global weather system, climate is represented as the outcome of biophysical process operating at a planetary scale that are captured through statistical measurement and whose dynamics are intelligible exclusively through sophisticated computer modelling (Hulme et al. 2008; Edwards 2010). On this basis, climate change is constructed in terms of statistical variations from enumerated baseline norms (IPCC 2007: 869). There are, of course, contrasting models of global climate systems, each of which is framed by different indices, parameters, baselines, timeframes and spatial boundaries that are culturally and politically influenced. The precise representation of physical climate, therefore, varies according to the professional conventions that govern what is measured and over what temporal periods and spatial zones. The socially mediated way that such models are put together, and the varied assumptions embedded within, create distinct results and this variability underscores a latent unease about the robustness of the predictions generated by climate models (Demeritt 2001; Hulme et al. 2008; Edwards 2010).

There is much to be gained in modelling atmospheric processes in this way. Notwithstanding inherent uncertainties concerning the accuracy of its predictive functions, modelling climate as a global system provides essential tools for both representing and understanding the changing biophysical relations between the earth's atmosphere, landmass and oceans. Contemporary climate science is unquestionably an incredibly powerful representational device and there would simply be no way to accurately envisage and simulate the relations between changes in solar and terrestrial radiation, ocean temperatures, precipitation, atmospheric and other processes without its evolving techniques. Through definitive advances in modelling geophysical processes both past and present, such analysis offer a fundamental contribution to our tenuous comprehension of atmospheric processes and climatic change. Climate science is, therefore, an essential means through which we can better understand our lived environments and the changes they experience. It has concurrently proved essential for making predictions about future climate scenarios (IPCC 2007). As David Demeritt highlighted in his careful interrogation of the politics and implicit biases of such models, "it is these powerful computer models that have been decisive in identifying the problem of future anthropogenic climate change and making it real for policy makers and the public" (2001: 309).

What is less noted, however, is the political impact of an intransigently rationalist idea of climate in which the latter is represented as an independent and external physical realm of global atmospheric processes that can be understood in abstraction from lived environments and their socio-ecological dynamics at various scales. The representational regime of climate science – with its rationalist division between society and nature – feeds directly into the biopolitical impetus to render climate change governable. Having fixed climate as something biophysical and that exists in separation from society, climate science can only bring the two back together as external influences. Humans are seen as an externality to climate, much as climate is seen as external to society. Each is viewed as part of an ontologically bounded system that corresponds to its own structuring processes and dynamics to which the other exerts an external influence in the form of impacts, constraints or shocks. On the one hand, the anthropogenic release of greenhouse gasses is seen to impact upon global climatic dynamics as an 'external forcing'. On the other, meteorological impacts are seen to shape human history in the form of constraints and shocks, most dramatically represented in terms of the projected impacts of climate change. Through this imagery of external influences, the clear separation of climate and society is enshrined and the ongoing choreography of social adaptation to natural climate can begin.

Such abstractions, of course, have a central place in climate science, where they are necessary simplifications for the purposes of modelling. My purpose, however, is not to discredit this form of scientific analysis but to question the conceptualisation of climate that it consolidates and circulates as a dominant yet silent presupposition of the adaptation discourse. The idea of an external climate as a force of nature that unfolds upon a similarly coherent society is now firmly engrained in the politics of mitigation and adaptation at an institutional level (Miller 2004; Hulme 2008). Even though the rationalist idea of climate is an abstraction from a multiplicity of diverse meteorological processes, it is discursively transformed into a coherent causal agent that emerges as an external threat to the present and seemingly fixed boundaries of societies. In so doing, an underlying physical reductionism has transcended the climate science literature and sets strong parameters for analyses that subsequently seek to connect this pre-externalised climate with the social world.

This form of representing climate in abstraction from society is not simply a consequence of an overbearing presence of climate science within governmental bodies of knowledge production such as the IPCC. As the following chapter argues in detail, the desire to extract and externalise climate from society is also rooted in a strong institutional urge to render climate change governable. By holding 'climate' and 'society' separate, with the former posed as an external threat to the latter, climate change can be represented not only as an outside disturbance to existing social orders but concurrently as an independent, external, knowable and therein potentially manageable phenomenon. This internal-external binary is key to creating of a 'world of adaptation' as a field of governmentality in which all aspects of society – from households to economies – are portrayed as necessarily adapting to external climatic stimuli, most dramatically instantiated in the form of 'hazards' such as drought, floods, storms, etc. Having suitably externalised and de-socialised climate, social dynamics can subsequently be brought into the analysis post facto as a context upon which climate sets to work. This is precisely the terrain on which the discourse and practices of adaptation take shape, including the grounds on which the concepts of vulnerability, adaptive capacity and resilience take form.

The IPCC reports, for example, follow a common mode of causal representation in which climatic change is presented in terms of changes to global biophysical processes that are then seen to impact downwards in the form of external shocks to social and environmental systems. The latter must subsequently adapt as best they can. Table 2.1, drawn from the IPCC, indicates how this representation involves a clear linearity of cause and effect: first, the climate changes, this then impacts prevailing environmental conditions, ultimately leaving societies needing to adapt. The overall effect, as Mike Hulme notes, is to further a methodology that extracts climate from "the complex interdependencies that shape human life" before positioning it in the role of "a dominant predictor variable" (2011: 247). Presently, the IPCC is attempting to address such critiques by more fully emphasising that social adaptations will impact upon how climatic impacts are experienced, yet this more reflective model still maintains a model of a cause-and-effect in which society is simply seen to exercise more control in shaping how climatic impacts will be manifested (see Carey 2014).

Humans and the production of climate

Although the rationalist foundations of adaptation are deeply ingrained in the dominant representations of climatic change, their dualistic separation between climate and society seems curiously misplaced given the way that human activities are inextricably embedded within atmospheric processes at various scales (see Head and Gibson 2012). Most emphatically, the concept of the 'anthropocene', highlighted by Chakrabarty above, indicates a planetary-scale zenith of this embeddedness that is inherent to the metabolism of industrial capitalism. It emphasises the degree to which humans now play a key role in co-producing climate as part of reproducing themselves on a day-to-day, year-to-year basis. Humans do not simply adapt to climatic change as some sort of external environmental stimulus. Rather, they are active protagonists in its production. The macro-scale dimensions of humanity's role in climatic production is, of course, most dramatically instantiated in this idea of the anthropocene in which global warming is the most dramatic instantiation. While such greenhouse gas emissions capture an essentially important facet of this mutual embeddedness within an industrial age, they do not capture the full scope of the historically intermeshing of social processes and climatic phenomena. From deforestation to large-scale irrigated agriculture, from urban design to industrial emissions, humans have actively co-produced climates at varied scales from the time they started to profoundly reshape their lived environments through settled agriculture (Foley et al. 2014).

Table 2.1 Examples of possible impacts of climate change due to changes in extreme weather and climate events, based on projections to the mid- to late-twenty-first century (IPCC 2007: 18)

Climatic Change		*Examples of major projected impacts by sector*			
Phenomenon and direction of trend	*Likelihood of future trends*	*Agriculture, forestry and ecosystems*	*Water resources*	*Human health*	*Industry, settlement and society*
Over most land areas, warmer and fewer cold days and nights, warmer and more frequent hot days and nights	Virtually certain	Increased yields in colder environments; decreased yields in warmer environments; increased insect outbreaks	Effects on water resources relying on snow melt; effects on some water supplies	Reduced human mortality from decreased cold exposure	Reduced energy demand for heating; increased demand for cooling; declining air quality in cities; reduced disruption to transport due to snow, ice; effects on winter tourism
Warm spells/heat waves. Frequency increases over most land areas	Very likely	Reduced yields in warmer regions due to heat stress; increased danger of wildfire	Increased water demand; water quality problems, e.g. algal blooms	Increased risk of heat-related mortality, especially for the elderly, chronically sick, very young and socially isolated	Reduction in quality of life for people in warm areas without appropriate housing; impacts on the elderly, very young and poor
Heavy precipitation events. Frequency increases over most areas	Very likely	Damage to crops; soil erosion, inability to cultivate land due to waterlogging of soils	Adverse effects on quality of surface and groundwater; contamination of water supply; water scarcity may be relieved	Increased risk of deaths, injuries and infectious, respiratory and skin diseases	Disruption of settlements, commerce, transport and societies due to flooding; pressures on urban and rural infrastructures; loss of property

In this respect, we can usefully situate the production of climate over a longer historical span and detach it from the prevailing idea of climate as inherently global in scope. In regard to the former, the work of climate scientist William Ruddiman has provocatively argued that the aggregated impacts of tree felling and the expansion of methane-producing farming activities since the advent of settled agriculture not only refashioned the way humans produced themselves but also simultaneously shaped meteorological processes on a planetary scale. As he lucidly puts it: "Before we built cities, before we invented writing, and before we founded the major religions, we were already altering climate. We were farming" (Ruddiman 2005: 4). In Ruddiman's interpretation, the slow yet accumulative impact of human-driven environmental engineering moderated prevailing physical climatic processes in ways that stalled the realisation of protracted periods of planetary cooling. While this approach usefully highlights the role of humans in co-producing climate as an essential element of historical process, it nonetheless retains the idea of climate as intrinsically global in scale. Ruddiman essentially backdates the anthropogenic impact upon a global climate system by some four millennia.

There is no need, however, to follow in the wake of climate science to view climate as something intrinsically unitary and global. Without doubt, atmospheric dynamics in which solar energy combines with hydrological processes in the circulation of air, water and heat across the earth's surface are fundamentally important determinants of meteorological phenomena in any given location. If we are to effectively capture its increasingly humanised dimensions, however, the idea of climate must incorporate more than global atmospheric processes. While the notion of the 'anthropocene' indicates an increase in the scale and scope of the co-production of climate between human and non-human forces, this is merely a current and expressly important instantiation of a longstanding process that occurs at various spatial scales. Deforestation, for example, not only affects 'global climate' by impacting upon the circulation of carbon dioxide and other gasses in the atmosphere. It also plays into the co-production of meteorological trends at micro, regional and macro levels through its impact on varied hydrological cycles. Analysts have mapped, for example, how deforestation in the Kenyan highlands creates a local daytime temperature differential of between 0.7 and 1.2 degrees centigrade between forested and cleared areas, with major implications for mosquito presence and breeding and therein human health (Afrane et al. 2006). At a more encompassing scale, changing forest covers in eastern Africa exert an impact upon monsoon rain patterns across the Indian Ocean into South Asia (Gupta, Thapliyal, Pal and Joshi 2005). Indeed, by asserting the importance of water cycling through humidity and trans-evaporation trends caused by forests, the much debated concept of the 'biotic pump' seeks to reinstate the centrality of dessicationism in affecting continental climatic trends. It challenges the scientific orthodoxy that temperature change is the primary driver of meteorological phenomena (Makarieva, Gorshkov and Li 2010, 2013).

While the overall significance of 'biotic pump' dynamics as a planetary climatic force continues to be disputed, the issue of deforestation nonetheless highlights the degree to which humans are prolific and tenacious agents of environmental transformation and, in so doing, become agents of climatic production at varied scales. By transforming the physical environment through changing forms of land use and management ranging from deforestation to mass-irrigation, humans shape meteorological phenomena from the local, to the regional and ultimately the global (Hoffman, Schroeder and Jackson 2003; Sacks et al. 2008). Beyond deforestation, for example, Sacks and collaborators model how large-scale irrigation changes levels of cloud cover and humidity with a significant alteration of regional temperatures (Sacks et al. 2008). More dramatically, Hebbert and Jankovic (2013) note how city design is unequivocally a form of anthropogenic climate change. In transforming the built environment through urbanisation, humans shape flows of air, heat and water, and they change temperatures and humidity levels. At the same time, yet unintentionally, the concentration of energy from urban industry, transport and other activities produces heat and air pollutants that also have a notable, yet uneven, impact upon the local climate and wider atmospheric dynamics at regional and global levels (see Erell, Pearlmutter and Williamson 2010). Industrial releases of 'black carbon', moreover, have been increasingly understood to impact upon regional climatic forces, with noted effects upon of surface and atmospheric temperatures, monsoon circulation and rainfall patterns (Ramanathan and Carmichael 2008).

While these processes of co-producing climate are largely unintentional, local-level climatic production also can be a deliberate strategy. Within agriculture, numerous techniques are purposely employed by farmers to produce amenable microclimates from the level of the individual field upwards. Planting or removing trees influences localised temperatures, wind velocity, evaporation and exposure to sunlight; while the intentional burning of straw or other waste materials is deployed to generate smog to trap outgoing heat radiation at night (Altieri 1995). Such practices are often intimately connected to locally embedded forms of knowledge. Michael Dove (2005), for example, notes how farmers in rain-fed north-western Pakistan have complex understandings of the interactions between trees and crops that focused on how trees shaded crops, regulating temperature and altering the soil moisture content with both good and bad consequences for yields and which necessitated localised management strategies. As I draw out further below, at this most micro of levels, local climates are formed through the interaction of meteorological forces, social practices and infrastructures, forms of knowledge and the role of other, non-human agents including trees, plants and animals. Producing climate is, therefore, an intrinsic part of producing the lived environment.

Beyond the climate-society dichotomy

Despite this intimate association between society and meteorological forces at varied scales, the primary conceptual lenses used to discuss them have remained trenchantly dualistic. The idea of adaptation, as noted above, emerges directly from a discursive process that separates out such intermeshed bundles of human and meteorological processes under the simple abstractions of 'climate' and 'society', which are, in turn, seen as parts of the 'natural' and 'social' worlds respectively (Head 2008). Once suitably cleansed of each other, these categories then relate as externals in which the bounded domain of climate impacts upon an equally bounded domain of society. It is on these grounds that we see the emergence of a language of shocks, stresses and ultimately, adaptation, as a means to bridge the tension between the two. We seem profoundly unwilling to relinquish a familiar set of Cartesian orientations that divide the world into 'natural' and 'social' domains despite the degree to which they appear markedly unsuited to capture the complex entanglements between social dynamics and meteorological processes (Moore 2013; and Chapter One). As a consequence, deep within the anthropocene, and its intimately interlaced socio-climatic processes, climate is still represented as something external to human energies to which the latter operate as an outside influence.

Notably, some authors within the framework of climate change adaptation do appear duly cognisant of the fragility of the nature–society dichotomy. Neil Adger, for example, argues that the notion of social-ecological systems, a concept inherited from the ecological resilience paradigm (Folke 2006), overcomes the duality by emphasising how "human action and social structures are integral to nature and hence any distinction between social and natural systems is arbitrary" (Adger 2006: 268). Despite this useful warning, however, Adger's ensuing analysis immediately replicates the binaries it purportedly seeks to challenge. Social-ecological systems are represented as a coupled pair that maintain their essential social and ecological natures but then engage in a series of mutual influences and feedbacks. This schematic maintains the dualism of systems that are essentially social or natural prior to their interactions, which are seen in terms of reciprocal influences and feedback loops. At the same time, climate itself is represented as an external source of shocks or stress to these otherwise coherent socio-ecological systems. On this basis, Adger is able to conclude that a "newly emerging interdisciplinary understanding of vulnerability and resilience demonstrates the co-evolutionary nature of social and natural systems – resilient ecosystems and resilient societies can better cope with external physical as well as sociopolitical stresses" (2006: 269). Despite its sound intentions, the familiar trope of independent yet mutually influencing 'social' and 'natural' systems that are rendered vulnerable through exogenous climatic factors appears too embedded within the discursive framework of climate change adaptation to shake loose (see Chapters Three and Four).

It is not accidental that this dichotomy between society and climate is so deeply engrained within the field of climate change adaptation. Discursively, it is precisely the prior ontological separation between climate and society that paves the way for their subsequent reconciliation under the rubric of adaptation. Adaptation emerges from this representational regime as the means to resolve the tensions between the two entities, wherein society adjusts to its natural environment by mediating threats and addressing vulnerabilities. To get beyond such static ontologies, it is necessary to bring production back into the picture. Societies do not simply 'exist' in distinction to nature wherein they go through a process of 'co-evolution' or 'adaptation' through mutual influences. Rather societies are actively produced as part of a lived environment that is indivisibly 'social' and 'natural'. From the agrarian environment to the urban conglomeration, lived environments are produced through constant transformative exchanges of energy and materials in which social relations are inextricably bound up with the biophysical forces that transform energy and materials in an active process of co-production (Kaika 2005; Swyngedouw 2006). It is only discursively that the results of this co-production between human and non-human forces are a posteriori encoded as 'society' and 'nature' through an act of boundary making aimed at rationalising and legitimating particular socio-ecological orders.

This raises a central conceptual question. If, as Adger contends, the distinction between society and nature that grounds the idea of adaptation is arbitrary, then why do we continue to act as if the separation was meaningful and build our frameworks on its skewed foundations? This is a vital issue because it pivots upon a key contradiction within the current discussion of climate change. As Leslie Head and Chris Gibson (2012) note, the vast body of literature on climatic change maintains dualistic ways of talking about contemporary socio-ecological transformations – including the language of human impacts, human interaction with environment, anthropogenic climate change, cultural landscapes, social-ecological systems – while the very processes they discuss confound such categorisation. As the previous chapter highlighted, the particular language we use to represent climatic change is a vitally important political issue because it serves to order and circumscribe the way we conceptualise contemporary socio-ecological changes and organise our actual and potential responses to them.

Putting culture in climate

A close engagement with politics of representing climate change is, therefore, a central preoccupation for critically orientated perspectives. Wary of the depoliticising implications of the rationalist account of abstract climate, for example, Fleming and Jankovic (2011) have proposed the need to rehabilitate the ancient Greek term '*Klima*' as a means to reconnect the notion of climate with a concrete sense of place and society. This impulse raises the key analytical question of how a more complete 're-socialisation' of climate might be accomplished. How might we understand our own creative involvement in producing both the climatic and social dimensions of the world that does not reduce such agency to a narrative of mutual impacts between ontologically distinct realms? One way forward, as Mike Hulme (2008, 2010) has proposed, is through the recognition of different meanings ascribed to climate across cultural contexts. As he evocatively expresses, the circuitry of climate modelling fundamentally detaches weather and climate from their human and cultural settings:

> A rainstorm which offers an African farmer the visceral experience of wind, dust, thunder, lightening, rain – and all the ensuing social, cultural and economic signifiers of these phenomena – is reduced to a number, say 17.8 mm.

> This number is propagated into the globalised and universalising machinery of meteorological and scientific institutions and assessments where it loses its identity.
>
> (Hulme 2008: 7)

Concerned with the political implications of this abstraction, Hulme advocates the need to complement the statistical representation of climate with an agency-centred one that recognises and validates contrasting cultural constructions of what weather and climate means in spatially and culturally specific locales over time (Hulme 2008; see also, Fleming and Jankovic 2011). Hulme's perspective gains support from anthropological studies of how weather is incorporated into 'traditional' cosmologies. Huber and Pedersen (1997), for example, highlight a distinctive rupture between a modernist framing of climate that rests on the idea of the environment as an ensemble of global, quantified interrelationships and 'non-modern' Tibetan knowledge that embeds weather as part of a system of local, qualitative interrelationships of humans, biophysical forces and spirit powers. For Hulme, the presence of diverse and contrasting conceptualisations create the potential for a tense yet productive amalgam of different constructions of climate in which "the scientific narrative of global climate change – and its regional manifestations – thus becomes entangled with the irrepressible personal experiences of local weather, whether these be traditionally proximate and sensuous experiences or newly vicarious and manufactured ones" (2010: 273).

Hulme's argument is well taken. Climatic phenomena are undoubtedly read and interpreted through contrasting frameworks that are grounded in specific social settings, practices and associated moral universes. Climate science itself could be considered exactly one such cultural frame produced from a historically embedded set of practices and bodies of knowledge (see, Ingold and Kurttila 2000). Indeed, if the idea of climate is discursively situated, as Hulme directs us towards, then there is no a priori reason to accept the Western scientific construction as a necessarily privileged one. For climate science, the practices involved in producing climate are designed precisely to separate it from the human world, to reduce it to an abstract set of measurements that are collected in ways that deliberately seek to avoid human influence so as to fix climate in a pure and unadulterated form. Human influence can then be brought in as an outside 'forcing' to an otherwise coherent model of atmospheric dynamics. This, as Hulme notes, is to produce climate in a way that strips it of tangible form and separates it precisely from how it is actually manifested and experienced.

On its own, however, this attempt to bring culture back into climate does not advance the critique of abstract climate quite far enough. It indicates that there can be differing cultural readings of an external climate and that climate – in the sense of cumulative and recurring patterns of weather over time within a given social space – is experienced and interpreted distinctly across cultural divides. The ways in which we attach meaning to the world around us, however, takes shape not in a separate cultural sphere in which an observer perceives a given outside world. Rather, they emerge through the practices by which we actively engage and reproduce ourselves in particular places and, in so doing, shape the world around us (Ingold 2011). The changing ideas of climate that Huber and Pederson noted in Tibet were not simply a clash between traditional and modern knowledge in which the latter gradually overcame the former. Rather, they emerged within the staggered transformation of the socio-ecology of the Tibetan plateau in which lives and livelihoods were slowly drawn into a new field of relations with different forms of political authority, organisations of labour, changing social hierarchies and new means of ordering the landscape (Clarke 1998; Yeh 2013). Changing ideas of climate, therefore, reflected deep changes in the production of lived environments and the associated livelihoods of their human populations.

So while the idea of climate seeks to capture the relatively durable patterns of weather that occur over time, such meteorological processes take tangible form within the broader socio-ecological relations in which they are situated. The visceral experience of rainstorms by an African farmer that Hulme raises is not simply an encounter between an isolated individual and a meteorological event. Rather, the meteorological phenomenon and the farmer relate to each other within the context of specific socio-ecological relations and infrastructures that shape how flows of water, wind, light and temperature take form within the landscape. The experience of a storm or a rainy season means something entirely different in urban Manhattan compared to rural Malawi, and this is not simply a question of cultural difference but of how meteorological forces are embedded within the broader built environment. Climates do not exist as an externality that people experience and interpret through an internal cultural frame. Rather, climates are produced through the tangible ways that meteorological forces are sourced into specific lived environments as active elements of a broad field of socio-ecological relations in which humans are actively involved.

Material climates

To put the social back in climate, therefore, is not merely a question of recognising different cultural interpretations of an objective external and natural climate. It is to emphasise how climates themselves are produced in part through the socio-ecological fixings that inhere meteorological forces within the active formation of the lived environment. This emphasis on active production of climate is captured by Tim Ingold in his concept of the inhabited 'weather-world' in which life actively reproduces itself through binding the weather into substantial, living forms (2007: 33; see also, Ingold 2011). For Ingold, these weather-worlds are indivisibly natural and social. They take shape across spatial scales at the intersection of social organisation, meteorological processes and the role of other biological actors. Although Ingold purposely avoids using the term climate, it is not necessary to surrender the term climate to the terrain of scientific rationalism. Instead, the socio-meteorological processes that he describes can be usefully considered as producing 'material climates'. This concept seeks to capture the fusion of meteorological forces, social organisation, physical infrastructure and discursive practices that shape the social and biophysical dimensions of our lived environments. In this framing, climate does not pre-exist as a natural environmental system that provides external stimuli to human lives. Rather, as Fleming and Jankovic note, climate takes tangible form in a 'hybrid realm' produced through the interactions of "land, water, air, living beings, people, and cultural institutions" (2011: 10). Material climates have no pristine existence in abstraction from socio-ecological processes and relationships. Rather, they come into being through the ways that meteorological forces are inhered within specific socio-ecological contexts. It is precisely through such processes that we experience climate as a tangible and meaningful dimension of human practice.

The notion of material climates, I believe, is central to re-framing the idea of climatic change in a way that can go beyond the dualistic conceptualisation of climate and society that characterise the field of climate change adaptation. What we term 'climate' does not exist outside society as a bounded external domain that serves as an environmental backdrop or constraint for social action and which buttresses society through exogenous shocks and stimuli (Figure 2.1). Rather, material climates are produced at various scales within the complex combinations of meteorological forces, social energies and other non-human agencies inherent to the production of lived environments (Figure 2.2). To suggest that humans play a role in producing climate at varied scales is not to argue for the primacy of social dynamics over non-human or biophysical forces. Rather, it is to emphasise that climate cannot be considered as something fundamentally external to social dynamics, which humans simply interpret and influence from the outside. We are fundamentally co-productive of material climates.

As schematised in Figures 2.1 and 2.2, the idea of co-production is fundamentally different from the notion of mutual influences between climate and society. While the latter emphasises a static ontology of climate and society locked into reciprocal impacts, the former emphasises instead the active production of material climates as part of the broader social and biophysical processes that produce lived environments. By interring meteorological processes into the production of our lived environments, humans do not simply employ an ontologically pre-formed 'climate' for our own purposes. In the process of bringing together

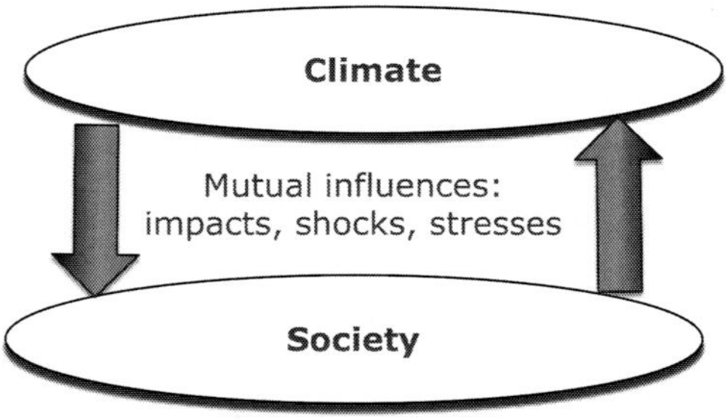

Figure 2.1 Climate and society as external mutual influences

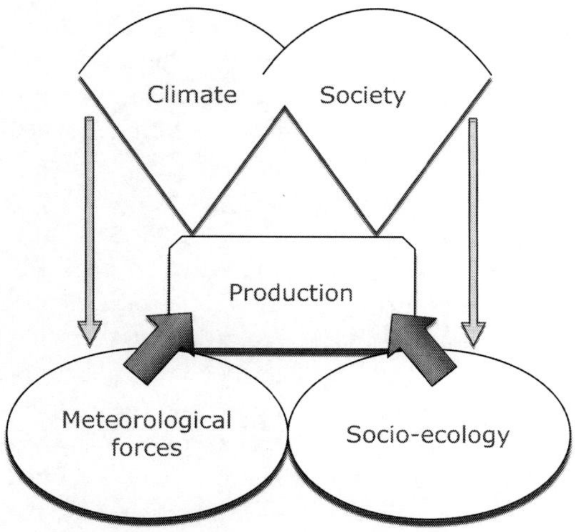

Figure 2.2 The co-production of climate and society

meteorological forces with forms of social organisation and built infrastructures, we create material climates as a tangible feature of the lived environment. At the same time, from global warming to urban microclimates, from deforestation to the release of industrial pollutants, humans are agents of climatic production at more encompassing scales.

By unpacking the climate–society dualism in this way, the concept of climate that animates the adaptation paradigm appears as a problematic basis for understanding the tangible effects of climatic change. The varied meteorological forces that are abstracted out into the notion of 'climate' are not encountered as abstract external elements of an exterior biophysical realm. Instead, they form essential elements of the tethered human and non-human processes through which our lived environments are produced. Humans play a key role in producing material climates. In producing ourselves we embed meteorological forces in our lived environments. We work with them, shape them and change them. In so doing, such meteorological forces shape us and often frustrate us. Occasionally they completely overwhelm us. In this manner, 'climate' is sometimes argued to exercise a form of agency, yet this agency is not that of an external force. It represents the power of meteorological forces that are inhered within the active production of specific lived environments. The 'natural' hazardousness attributed to climate is nothing less than the inverse side of its social productiveness within a specific socio-ecological setting. This point, however, is repeatedly lost within confines of the climate change adaptation framework and the notion of climate as external stimuli, shocks and stresses.

It may be counterintuitive, but one implication of this reframing is that climate change is not solely about biophysical changes to a global climate system and human reactions to them. Rather, a change in material climate can be produced through re-ordering the socio-ecological relations at regionalised or local levels that affects the way that meteorological forces are inhered within the landscape. Consider, for example, an agrarian community undergoing a shift driven by colonial duress from pastoral practices on common rain-fed lands to a form of settled agriculture with privatised property and canalised irrigation (see Chapter Five). Through such socio-ecological ruptures, both the material expressions and lived experiences of climatic processes are profoundly transformed. Under this shift, temperature, humidity, rainfall, and variations in light and wind are radically reworked as core elements of the lived environment. Such physical processes now express themselves in and through new forms of socio-ecological organisation that facilitate flows of water, types of cropping, vegetation and tree cover, the production and movement of different kinds of labouring bodies within the agrarian environment and the disbursement or destruction of particular forms of knowledge. Under this transformation, the same physical determinants of enumerated abstract climate that can be produced though statistical indexing are manifested and experienced entirely differently according to how they are reworked as productive elements of a lived environment. Even as the same physical patterns of rainfall and temperature may persist in statistical records, climate comes into being as something radically different. It has irrevocably changed.

A change in climate, therefore, is not simply an alteration to patterns of select meteorological variables in an external environment shaped by biophysical forces operating at a global level. It can also express a reordering of the socio-ecological relations through which humans seek to reproduce themselves and, in so doing, co-produce material climates. From this perspective, we can better understand why North American colonists in the seventeenth and eighteenth centuries repeatedly and stubbornly claimed that they were changing and improving their local climate by reordering localised socio-ecological relations through deforestation, enclosure and settled agriculture (Vogel 2011). Although the meteorological record might tentatively indicate that there was little significant change in average temperatures, the socio-ecological transformation of the landscape did indeed produce a tangibly different material climate, which colonial settlers insisted served to temper winters and render the setting less 'unhealthy'. Of course, the indigenous inhabitants of such lands had quite different interpretations of the nature and virtue of such imposed climatic change. As I address below, the production of material climates is inherently shaped by the active presence of social hierarchies and forms of power operating across spatial scales.

The power of producing climate

Reframing climate in this way changes the fundamental questions we face in an era in which the co-production of climate has become expansive and increasingly volatile. Humans do not simply adapt *to* climate change; they co-produce climates in ongoing and unequal ways. If climatic change is not something that simply occurs 'out there' and subsequently impacts down upon society but rather is engrained within the ongoing production of lived environments across geographic scales, then the discourse of climate change adaptation must be questioned for the way it marginalises the co-production of climate and its associated socio-ecological dynamics from its analytical coordinates. Most fundamentally, grappling with the co-production of climate prefigures a transition from thinking about 'adapting to climate change' towards figuring out how we must co-produce climate differently. To do so, production – in the encompassing sense of the collective socio-ecological metabolisms through which we reproduce ourselves over time – needs to be brought back into the heart of the climate change debate. This entails moving beyond the fetishised confines of the adaptation/mitigation dichotomy that structures present debates. It involves writing climate into our historical narratives of power and contestation, co-operation and conflict not as an outside influence upon them but as an integral dynamic element of them (Moore 2013).

Once subjected to this 'socialisation', climate emerges as a deeply historical process. Material climates are actively made and unmade at changing scales through the ceaseless interaction of meteorological processes, human productive activities and other non-human dynamics that bind together varied agents from the 'biotic pump' of tropical forests to the methane-laden farts of agro-industrial cattle herds. It is, of course, possible to claim that climate still has an objective external existence independent of these socio-ecological fixings, as captured in the statistical variables of meteorologists and modelled by climate science. The idea of material climates does not dismiss the rationalist representation of climate as an abstraction. Instead, it points to the poverty of an analysis of climatic change that follows from the notion of an external, pre-formed abstract climate that has 'impacts' upon the social world. Despite its centrality to the creation of predictive global climate models, such representations are precisely built upon abstractions from the grounded socio-ecological processes and infrastructures through which climate is produced as something visceral and meaningful, productive and destructive. It is therefore profoundly troubling that rationalist renderings of climate are quietly dominant within the field of climate change adaptation. As several critics have noted, the overall effect is a pervasive climate fetishism in which the abstract category of 'climate' is repeatedly transformed into a master causal agent of human futures (Swyngedouw 2010; Fleming and Jankovic 2011; Hulme 2011).

In contrast, by socialising climate it is possible to fundamentally challenge the accepted causal drivers and spatial registers that frame the discursive parameters of climate change adaptation. The notion of material climates disputes the idea that climatic change is simply an external biophysical process driven at a global scale that then refracts downwards to lower levels, necessitating a process of adaptation. Rather, material climates are co-produced at the intersection of human agency and biophysical processes at varied spatial scales. This is not to dismiss the central importance of what is termed 'global climate change'. The escalating anthropogenic emission of greenhouse gasses is unequivocally a vital element of contemporary climatic production. Rather, the purpose is to situate processes of climatic change within a far broader spectrum of socio-ecological transformations of which they form an essential part. The production of climate is emphatically a multi-scalar process. Climatic change occurs not only through greenhouse gas emissions that affect solar radiation in the high atmosphere but also by processes occurring at local and regional levels that reshape lived environments and the role and character of meteorological forces within it. Climatic change, therefore, is not uniquely global and external. It is actively pursued and contested by social actors at a range of scales with intertwined social and biophysical determinants.

This reframing of climate strikingly challenges how we conceptualise agency within climatic change. By situating processes of climatic production across scales and driven by human and more-than-human agencies, we can better conceptualise how the production of climate is interlaced within complex power relations. Varied actors seek to shape the production of lived environments in ways that provide benefits and security to them and externalise the detrimental outcomes of such processes onto others. Such power relations are frequently written into the physical forms of the lived environment. As David Mosse notes in his study of irrigation within agrarian Tamil Nadu, for example, local elites ensure that unequal systems of water sourcing are built into the physical designs of the agrarian environment, including the layout of field irrigation channels and the selective disrepair of specific well structures. Social and biophysical forces therein become inseparable agents in the production of a vastly unequal lived environment and its material climate. On the one hand, the social relations of control over land become expressed through meteorological forces. The farmer deprived of irrigated fields through historically unequal property rights becomes heavily reliant on the fickleness of seasonal rains. On the other, meteorological forces are expressed through social structures. Harvest failure through prolonged drought manifests itself in the form of escalating indebtedness to moneylenders and subsequent relationships of dependence and exploitation (see Chapters Six and Seven). Material climates, therefore, are simultaneously expressive and productive of ingrained human inequalities.

While social agents seek in diverse ways to shape and contest the production of lived environments in both their social and climatic dimensions, they do not do so in conditions of their own choosing. The seemingly localised dynamics of material climates can only be understood within the context of socio-ecological processes that are stretched across space and time. While humans have played a role in co-producing climate on various scales since the advent of settled agriculture, it is evidently since the emergence of the capitalist era – and specifically industrial capitalism – that the scope of scale of climatic co-production has risen to entirely unparalleled degree (Clark and York 2005; Chakrabarty 2009; Foley et al. 2014). As such, human actions are inevitably and deeply conditioned by the broader socio-ecological dynamics of capital accumulation, in which social agents experience strong impulses to drive forward both the scale and scope of metabolic activities that profoundly transform the lived environment and yet displace the associated costs onto others (Smith 1984; Harvey 1996; Swyngedouw 2006). The production of material climates, therefore, forms part of a global division of production and consumption that is profoundly uneven and has been driven by new and complex logics with distinctly uneven outcomes (Taylor 2009).

What must not be lost, however, is that while the exponential increase in greenhouse gas emissions produced by industrial activity is of central importance to processes of contemporary socio-ecological transformation, the complex agencies associated with capitalism have also transformed climate at other scales and along different temporal axes. Indeed, the history of capitalism is one of the constant production and transformation of lived environments and material climates, with both planned and unplanned consequences. These have been unevenly experienced across spatial divides, temporal frames and the social divides of class, gender and other fractures. We might take note, therein, of how the frequently bloody transformations that attended the rise of capitalism, and which continue to mark its present, were closely bound into the production of climate. The commodification of land as private property, the rise of plantation agriculture and slavery, waves of deforestation, massive projects of engineering and displacement, the mechanisation and commodification of agriculture and the dramatic increase in urbanity that continues to accelerate in the present have all been processes of landscape transformation and climatic production (Hornborg 2001; Moore 2010). The reworking of material climates on an expanding scale, therefore, is intrinsic to the workings of capitalism and must be understood within this context. Such dimensions of contemporary climatic production, as I examine in the following chapter, remains strongly marginalised within the discourse of climate change adaptation.

Notes

1 As noted below, important exceptions include Demeritt (2001), Head (2008), Hulme (2008, 2011), Fleming and Jankovic (2011), Heymann (2011), Head and Gibson (2012), Moore (2013). The clear influence of these authors is perceptible throughout this chapter.

2 As Carey (2011) notes, theories of climate as a determining factor of human capacity that defined levels of civilisation were produced well into the twentieth century and led to stereotypically negative depictions of tropical inhabitants. Jared Diamond, perhaps, is just the most recent and prominent incarnation of this trend.

References

Adger, W. N. 2006. "Vulnerability." *Global Environmental Change* 16(3): 268–81.

Afrane, Y., G. Zhou, B. Lawson, A. Githeko, and G. Yan. 2006. "Effects of microclimatic changes caused by deforestation on the survivorship and reproductive fitness of anopheles gambiae in western Kenya highlands." *American Journal of Tropical Medicine and Hygiene* 74(5): 772–78.

Altieri, M. A. 1995. *Agroecology: The Science of Sustainable Agriculture*. Boulder, CO: Westview Press.

Anderson, K. 2005. *Predicting the Weather: Victorians and the Science of Meteorology*. Chicago: University of Chicago Press.

Carey, M. 2011. "Inventing Caribbean climates: How science, medicine and tourism changed tropical weather from deadly to healthy." *Osiris* 26(1): 129–41.

Carey, M. 2012. "Climate and history: a critical review of historical climatology and climate change historiography." *WIREs Climate Change* 3(3): 233–49.

Carey, M. 2014. "Science, Models, and Historians: Toward a Critical Climate History." *Environmental History* forthcoming.

Chakrabarty, D. 2009. "The climate of history: Four theses." *Critical Inquiry* 35(4): 197–226.

Clark, B., and R. York. 2005. "Carbon metabolism: Global capitalism, climate change, and the biospheric rift." *Theory and Society* 34(4): 391–428.

Clarke, G. 1998. "Development, Society and Environment in Tibet." Pp. 1–43 in *Proceedings of the 7th Seminar of the International Association for Tibetan Studies*, edited by Ernst Steinkellner. Wien: Osterreichischen Akademie der Wissenschaften.

Demeritt, D. 2001. "The construction of global warming and the politics of science." *Annals of the Association of American Geographers* 91(2): 307–37.

Dove, M. 2005. "Shade: Throwing Light on Politics and Ecology in Contemporary Pakistan." Pp. 217–55 in *Political Ecology across Spaces, Scales, and Social Groups*, edited by Susan Paulson and Lisa Gezon. New Brunswick, NJ: Rutgers Press.

Edwards, P. N. 2010. *A Vast Machine: Computer Models, Climate Data, and the Politics of Global Warming*. Cambridge, MA: MIT Press.

Erell, E., D. Pearlmutter, and T. Williamson. 2010. *Urban Microclimate*. London: Routledge.

Fleming, J.R. 1998. *Historical Perspectives on Climate Change*. Oxford: Oxford University Press.

Fleming, J. R., and V. Jankovic. 2011. "Revisiting Klima." *Osiris* 26(1): 1–15.

Foley, S., D. Gronenborn, M. Andreae, J. Kadereit, J. Esper, and D. Scholz. 2014. "The Palaeoanthropocene – The beginnings of anthropogenic environmental change." *Anthropocene* 3(1): 1–6.

Folke, C. 2006. "Resilience: The emergence of a perspective for social-ecological systems analyses." *Global Environmental Change* 16(4): 253–67.

Gidwani, V. 2008. *Capital, Interrupted*. Minneapolis: University of Minnesota Press.

Golinsky, J. 2008. "American Climate and the Civilization of Nature." Pp. 151–74 in *Science and Empire in the Atlantic World*, edited by James Delbourgo and Nicholas Dew. New York: Routledge.

Grove, R. 1997. *Ecology, Climate and Empire: Colonialism and Global Environmental History, 1400–1940*. Cambridge: White Horse Press.

Gupta, A., P. K. Thapliyal, P. K. Pal, and P. C. Joshi. 2005. "Impact of deforestation on Indian monsoon – A GCM sensitivity study." *Journal of Indian Geophysical Union* 9(2): 97–104.

Harvey, D. 1996. *Justice, Nature and the Geography of Difference*. Oxford: Blackwell.

Hazareesingh, S. 2012. "Cotton, climate and colonialism in Dharwar, western India, 1840–1880." *Journal of Historical Geography* 38(1): 1–17.

Head, L. 2008. "Is the concept of human impacts past its use-by date?" *The Holocene* 18(3): 373–77.

Head, L., and C. Gibson. 2012. "Becoming differently modern: Geographic contributions to a generative climate politics." *Progress in Human Geography* 36(6): 699–714.

Hebbert, M., and V. Jankovic. 2013. "Cities and climate change: The precedents and why they matter." *Urban Studies* 50(7): 1332–47.

Heymann, M. 2011. "The evolution of climate ideas and knowledge." *WIREs Climate Change* 1(1): 581–98.

Hoffman, W., W. Schroeder, and R. Jackson. 2003. "Regional feedbacks among fire, climate and tropical deforestation." *Journal of Geophysical Research* 108(D23): 4721–32.

Hornborg, A. 2001. *The Power of the Machine: Global Inequalities of Economy, Technology, and Environment*. Walnut Creek, CA: AltaMira Press.

Huber, T., and P. Pedersen. 1997. "Meteorological knowledge and environmental ideas in traditional and modern societies: The case of Tibet." *Journal of the Royal Anthropological Institute* 3(3): 577–98.

Hulme, M. 2008. "Geographical work at the boundaries of climate change." *Transactions of the Institute of British Geographers* 33(1): 5–11.

Hulme, M. 2010. "Cosmopolitan Climates: Hybridity, foresight and meaning." *Theory, Culture and Society* 27(2–3): 267–76.

Hulme, M. 2011. "Reducing the future to climate: A story of climate determinism and reductionism." *Kima* 26(1): 245–66.

Hulme, M., S. Dessai, I. Lorenzoni, and D. Nelson. 2008. "Unstable climates: Exploring the statistical and social constructions of 'normal' climate." *Geoforum* 40(2): 197–206.

Ingold, T. 2007. "Earth, sky, wind, and weather." *Journal of the Royal Anthropological Institute* 13(S1): S19–S38.

Ingold, T. 2011. *Being Alive: Essays on Movement, Knowledge and Description*. London: Routledge.

Ingold, T., and T. Kurttila. 2000. "Perceiving the environment in Finnish Lapland." *Body and Society* 6(1): 183–98.

IPCC. 2007. *Climate Change 2007: The Scientific Basis. WG I contribution to IPCC 4th assessment report*. Cambridge: Cambridge University Press.

Jankovic, V. 2010. "Climates as Commodities: Jean Pierre Purry and the modelling of the best climate on Earth." *Studies in History and Philosophy of Modern Physics* 41(2): 201–07.

Kaika, M. 2005. *City of Flows: Modernity, Nature and the City*. London: Routledge.

Makarieva, A., V. Gorshkov, and B. Li. 2010. "The biotic pump: Condensation, atmospheric dynamics and climate." *International Journal of Water* 5(4): 365–85.

Makarieva, A., V. Gorshkov, and B. Li. 2013. "Revisiting forest impact on atmospheric water vapor transport and precipitation." *Theoretical and Applied Climatology* 111(1): 79–96.

Miller, C. 2004. "Climate Science and the Making of a Global Political Order." Pp. 46–66 in *States of Knowledge: The Co-Production of Science and the Social Order*, edited by S. Jasanoff. London: Routledge.

Mitchell, T. 2002. *Rule of Experts: Egypt, Techno-Politics, Modernity*. Berkeley: University of California Press.

Moore, J. 2010. "'Amsterdam is standing on Norway' part II: The global North Atlantic in the ecological revolution of the long seventeenth century." *Journal of Agrarian Change* 10(2): 188–227.

Moore, J. 2013. "From object to Oikeios: Environment-making in the capitalist world-ecology." Unpublished paper. Department of Sociology, Binghamton University. http://www.jasonwmoore.com/uploads/Moore__From_Object_to_Oikeios__for_website__May_2013.pdf accessed October 1st, 2013.

Prakash, G. 1999. *Another Reason: Science and the Imagination of Modern India*. Princeton, NJ: Princeton University Press.

Ramanathan, V., and G. Carmichael. 2008. "Global and regional climate changes due to black carbon." *Nature Geosciences* 1(4): 221–27.

Ruddiman, W. F. 2005. *Plows, Plagues, and Petroleum: How Humans Took Control of Climate*. Princeton, NJ: Princeton University Press.

Sacks, W., B. Cook, N. Buenning, S. Levis, and J. Helkowski. 2008. "Effects of global irrigation on the near-surface climate." *NASA Publications* 7–23–2008 Paper 5: 1–18.

Smith, N. 1984. *Uneven Development: Nature, Capital, and the Production of Space*. New York: Blackwell.

Swyngedouw, E. 2006. "Circulations and metabolisms: (Hybrid) natures and (cyborg) cities." *Science as Culture* 15(2): 105–21.

Swyngedouw, E. 2010. "Apocalypse forever? Post-political populism and the spectre of climate change." *Theory, Culture and Society* 27(2–3): 213–32.

Taylor, M. 2009. "Displacing insecurity in a divided world: Global security, international development and the endless accumulation of capital." *Third World Quarterly* 30(1): 147–62.

Vogel, B. 2011. "The letter from Dublin: Climate change, colonialism, and the royal society in the seventeenth century." *Osiris* 26(1): 111–28.

Williams, R. 1980. *Problems in Materialism and Culture*. London: Verso.

Yeh, E. T. 2013. *Taming Tibet: Landscape Transformation and the Gift of Chinese Development*. Ithaca, NY: Cornell University Press.

3 Making a world of adaptation

On May 9, 2013, a symbolic threshold was passed. For the first time since the mid-Pliocene some two to four million years ago, the concentration of carbon dioxide in the earth's atmosphere rose above 400 ppm. One of the implications of this event was that it demonstrated how climate change mitigation – the attempt to reduce the magnitude and rate of climate change by restricting greenhouse gas emissions – has been an unqualified failure. Owing to the intransigence of primarily (but not exclusively) the industrialised countries of the developed West, the reduction of CO2 emissions and other greenhouse gasses will not be rapid enough in timeframe or sufficient enough in scale to avoid significant changes to temperatures and climatic cycles on a world scale. As a result, the globe is set to continue experiencing a secular trend of rising temperatures leading to increased climatic variability and a growing frequency of weather extremes. In the words of historian Dipesh Chakrabarty, "[h]umans, collectively, now have an agency in determining the climate of the planet as a whole, a privilege reserved in the past only for very largescale geophysical forces" (2012: 9). In these terms, we have entered an age labelled the 'anthropocene' in which the outcomes of collective human activity act as a defining geological force of climatic transformation.

Although debate continues over the precise impacts of such shifts, there exists a broad consensus that the social and environmental effects of climatic change will be both extensive and unevenly distributed. On the one hand, the negative impacts will be experienced most severely in the poorest countries of the world, which contributed least to the problem. On the other, on top of this unevenness, the adverse impacts of such processes are likely to impact most intensely among populations that already face significant levels of poverty and vulnerability therein compounding their marginalisation (Tol et al. 2004; Thomas and Twyman 2005; Adger 2006a; Stern 2007). In particular, rural inhabitants in much of Asia, Africa and Latin America are perceived as uniquely exposed to climatic variability owing to a widespread dependence upon natural resource-based livelihoods, insufficient public infrastructure and historically entrenched poverty (United Nations Framework Convention on Climate Change 2007; World Bank 2008; Verner 2010). In this context, there are manifold fears that increased water stress, decreased agricultural yields, reduced returns from livestock and augmented pressure on existing resources will translate into significant social distress

(Fisher et al. 2005; Morton 2007). At the same time, infrastructural weaknesses and the paucity of public services in such regions are argued to leave such populations greatly exposed to climate change-related hazards, from droughts to heat waves to floods (Agrawal and Perrin 2008).

In the context of such anticipated impacts, the spectre of anthropogenic climate change has unequivocally cast a shadow over the concerns and practices commonly collected under the rubric of development. The UNDP, for example, suggests that climate change contains the potential to cause catastrophic impacts upon human welfare that "calls into question the Enlightenment principle that human progress will make the future look better than the past" (UNDP 2007: 1). This claim is made on the basis of estimates that over the next half century climate change impacts ranging from agricultural disruption to rising sea levels are likely to lead to an extra 600 million people experiencing malnutrition, an additional 1.8 billion people living in water-stressed environments and 330 million people permanently displaced (UNDP 2007: 9). Given the magnitude of such impacts, many have suggested that climate change threatens to disrupt or render obsolete existing ways of 'doing development'. As Boyd et al. suggest: "Development futures are already unclear and difficult to plan, even before adding the trump of the uncertainty of climate change into the mix. Bringing the two together coherently is an unprecedented challenge" (2009: 660).

As a response, international institutions, national governments and NGOs have overwhelmingly gravitated towards the concept of climate change adaptation, a move that was formalised in the 2006 UN Nairobi agreement. In descriptive terms, climate change adaptation is commonly defined as a process of adjustment of social, environmental and economic systems so as to alleviate the actual and anticipated adverse effects of climate change and to take advantage of new opportunities (IPCC 2001). The rise of climate change adaptation has been rapid and comprehensive. Adaptation now pervades a litany of development interventions, and all major international development institutions from the UNDP to the Organisation for Economic Cooperation and Development (OECD) project the need to support and integrate climate change adaptation as a key development policy goal. The World Bank, for example, began to integrate climate change adaptation into its operations in the mid-2000s. Alongside creating new research and lending facilities specifically aimed at supporting adaptation projects, the bank simultaneously promoted the full integration of adaptation goals across its existing lending programme. As a result, all its current and future development projects are projected to address climate change adaptation as integral parts of their design and operation. The aim of this 'climate risk management', in the World Bank's view, is to make existing development investments more resilient to climate variability and extreme weather events while simultaneously improving the impact of development efforts in the present (World Bank 2006a).

In foregrounding a synthesis of climate change adaptation with development programming, the World Bank is far from unique. The goal of climate change adaptation has been mainstreamed into a full spectrum of development strategies and projects across institutions, a process that underscores its dramatic ascendance within the field of international development (Huq and Reid 2004; Burton et al. 2007; Swart and Raes 2007; Bizikova et al. 2010). Similarly, national governments across the globe have unvaryingly produced climate change adaptation programmes that are projected as central elements of sustainable development strategies. A mutually reinforcing logic underscores this process: without climate change adaptation, the gains of development could be lost; without appropriate development, societies remain highly vulnerable to the impacts of climate change (IPCC 2007; OECD 2009; World Bank 2009). As the IPCC puts it, "sustainable development can reduce vulnerability to climate change, and climate change could impede nations' abilities to achieve sustainable development pathways" (2007: 20). On the basis of this convergence, climate change adaptation and its core underlying concepts – vulnerability, resilience, and adaptive capacity – now assume a place within the lexicon of mainstream development thinking alongside competiveness, good governance, empowerment and sustainability. The goal of adaptation is therein deployed across a range of institutional sites as a basis on which to design and legitimise development policy and other interventions (OECD 2009; World Bank 2009).

The significance of this emergence of climate change adaptation as a core paradigm within the field of international development should not be underestimated. In the context of powerful imageries of the projected impacts of climatic change upon human societies, adaptation stands simultaneously as a normative goal and as a framework for understanding for the intersection of climate, environment and society. The framework of climate change adaptation is therein reshaping thinking about conjoined social and environmental change on a global scale. It reconfigures prevailing categories of analysis, influences what gets measured and determines how resulting data gets interpreted. At the same time, it alters the priorities and possible actions for a range of actors, from governmental bodies, to international development institutions, to NGOs, to community-based groups. New claims relating to development priorities, the use of resources and the legitimacy to act are now increasingly being made within the framework of climate change adaptation; while existing claims are simultaneously repackaged under the adaptation rubric (Ireland 2012).

The politics of adaptation

Given this role in setting priorities and determining the distribution of public and private resources, the framework of climate change adaptation is inherently political. It shapes how claims can be made, who can speak with authority, who can produce knowledge on behalf of whom, who is funded and entrusted with solving the issues that it identifies and who should live with consequences of the issues it marginalises. Yet, despite the intrinsically political nature of both the discourse and practices of climate change adaptation, much of the policy and academic literature on adaptation obscures this point. Adaptation is presented as a necessity that stands outside of politics. Part of the problem is that adaptation seems a natural and inevitable response to change in our lived environments. We are struck by the seeming self-evidence of the notion that humans must adapt to the challenge posed by environmental change in order to survive. The concept of adaptation therein resonates with deep-seated ideas of natural evolutionary trajectories and historical change (Orlove 2009). Indeed, influential paradigms within the discourse of climate change adaptation, such as the resilience perspective, purposely draw upon natural analogies of adaptation to external stresses in order to understand contemporary societies in change. To question adaptation, it seems, would be tantamount to resisting nature itself.

And yet, while it is a truism that humans perpetually change alongside their social and ecological settings, the concept of adaptation to climate change does not pre-exist the discursive frameworks that produce it as an analytical object and a site of development intervention. As the following chapters argue, the normalisation of the notion of climate change adaptation at a governmental level and its actualisation through particular policies, institution building and identity formation serve to create strong framing boundaries around how we conceptualise climatic change and societal transformation. This is to argue that climate change adaptation cannot be considered a neutral framework that simply needs filling with contents that can be more or less encompassing in scope and more or less progressive in political terms. Rather, the framework already contains a number of core assumptions that condition the way in which we envisage the linkages between social and environmental change. In so doing, the language of adaptation legitimates particular forms of knowledge, governance and policy, yet it does so silently without making explicit the analytical and normative assumptions that are built into its conceptual foundations. In many cases, critical approaches to climate change adaptation rail against these constraints and seek to overcome them within the bounds of the adaptation paradigm (e.g. Adger, Lorenzoni and O'Brien 2010; Pelling 2011). Such attempts, as I later argue, are riveted with tensions stemming from the attempt to radicalise a paradigm that is fundamentally constraining in its foundational assumptions.

It is, therefore, useful to step back from the compelling rhetoric of 'adaptation now!' in order to consider climate change adaptation not as a self-evident body of knowledge and practices but, instead, as a discourse: that is, a set of relations between forms of knowledge, structures of power, institutional practices and prevailing technologies that delineate ways of thinking about and acting upon processes of social and ecological change (Escobar 1995; Mitchell 2002). To pose adaptation as a discourse is not first and foremost to critique existing ways of actualising adaptation in policy terms, although this is certainly one consequence. More fundamentally, it is to problematise what it means to interpret and act upon the world through the lens of climate change adaptation. To do so is to consider how the language and practices of adaptation produce the idea of climatic change in ways that are attuned to particular strategies of governance and rule. The point of de-framing climate change adaptation in this way is to uncover and challenge the political underpinnings that are naturalised within its discursive frame.

It might immediately be countered that the field of climate change adaptation is simply too diverse and heterogeneous to be identified in terms of a singular discursive framework. Numerous and contrasting analytical perspectives populate its intellectual horizon and they tend towards diverging interpretations of both the politics and policies of adaptation (see Pelling 2011 for a comprehensive overview). As I elaborate below, however, while such perspectives debate the aims and processes of adaptation, they nonetheless share a common set of conceptual foundations. Together these have led to a pervasive standardisation over how climate change adaptation is represented and enacted in both governmental and academic fields (Bassett and Fogelman 2013). Such convergence is most strongly reflected in the coherence of the adaptation concept within international policymaking, where it now operates as a largely unquestioned conceptual apparatus through which to represent climatic change and its social impacts. Even a casual glance at the multiplicity of recent climate change adaptation strategies, from countries as diverse as Mauritius to Mongolia, demonstrate a remarkable cohesion in their underlying concepts, frameworks and their operationalisation. Such National Adaptation Plans of Action invariably embrace a threefold analytical framework based on the core notions of vulnerability, resilience and adaptive capacity.

Box 3.1 The 'holy trinity' of climate change adaptation (IPCC, 2007)

Vulnerability: the degree to which a system is susceptible to, and unable to cope with, adverse effects of climate change, including climate variability and extremes. Vulnerability is a function of the character, magnitude and rate of climate change and variation to which a system is exposed, its sensitivity and its adaptive capacity.

Adaptive capacity: the ability of a system to adjust to climate change (including climate variability and extremes) to moderate potential damages, to take advantage of opportunities or to cope with the consequences of change.

Resilience: the ability of a social or ecological system to absorb disturbances while retaining the same basic structure and ways of functioning, the capacity for self-organisation and the capacity to adapt naturally to stress and change.

These concepts form the analytical core of the climate change adaptation framework. They have become the standardised and almost exclusive means for conceptualising climatic change and associated social and ecological transformations. Their conceptual range, moreover, is projected to be universal. For example, the Nepalese government published its 2011 action plan for dealing with climate change explicitly stipulating how "the framework proposed and the tools and approaches recommended draw from current understandings of global and national climate adaptation." The document then proceeds to discuss the country's vulnerabilities to climate change before projecting means to enhance the "adaptive capacity of communities through resilient development plans"

(Government of Nepal 2011). Elsewhere in the region, the same series of concepts orientate the equivalent strategies for Pakistan (see Chapter Six), India (Chapter Seven), and Bangladesh (Ministry of Environment and Forests, 2008). This common conceptual orientation occurs despite the vastly different nature of the challenges each country faces and the diversity of socio-ecological settings within and across these settings. In each case, local context is used to fill out the pre-formed categories of vulnerability and resilience through which the discursive framework of climate change frames the issues.

Such homogeneity reflects the way in which national governments draw upon the now copious literature produced by the IPCC, World Bank, OECD, UNDP and others that reproduces this framework. As J.C. Gaillard (2010) notes, the summary for policy makers of the IPCC report on 'Impacts, Adaptation and Vulnerability', used the concept of vulnerability 30 times, adaptive capacity 25 times, and resilience four times within only 16 pages (see IPCC 2007). Within the policy discourse surrounding climate change adaptation, these concepts are so deeply embedded that they are sometimes presented as self-evident, stemming directly from the nature of the threat posed by climatic change. "The case for resilience has never been stronger," proclaims Ban Ki-moon, secretary-general of the UN, in his preface to the World Bank report 'Turn Down the Heat: Climate Extremes, Regional Impacts and the Case for Resilience' (World Bank 2013). Apparently, such is the self-evidence of the concept that this lauded and lengthy report never defines what is meant by the 'resilience' for which it is making a case. A world of climatic disturbance is simply assumed to evoke the need for 'resilient' societies and ecosystems.[1]

In the wake of this institutionalisation, climate change adaptation has become a universalising framework predicated upon an underlying set of 'travelling rationalities' (Mosse 2011). The latter term refers to concepts produced and circulated within international agencies as set of seemingly universal concepts that transcend political, economic, ecological and cultural settings. Now ubiquitous throughout both the policy and academic literature, the trio of vulnerability, adaptive capacity and resilience operate precisely in this fashion. On a practical level, they operate as a set of 'plug and play' concepts that can be imported into any given setting to rationalise standardised policy planning across diverse socio-ecological contexts. Consider, for example, the way that Ensor and Berger note in their review of community-based adaptation across eight heterogeneous settings, that the "principal adaptation activities are identified as vulnerability reduction, building adaptive capacity and strengthening resilience" (2009: 6). In this case, community-based adaptation is seen as a means to provide local content to fill out these universal concepts that themselves precede the historical particularities of any given location and its underlying socio-ecological dynamics.

Such formulaic renditions of adaptation, it could be countered, are merely a function of international policymaking's dual purpose of public knowledge dissemination and policy coordination, both of which inherently tend towards the simplification of complex issues. While the international policy literature indisputably betrays a strong standardisation of concepts, the level of academic research is far more diverse. In the growing body of literature on adaptation there exists a ferment of knowledge production that is replete with contradiction and contestation. This is, of course, true to a point. Much of this academic work examines diverging means of conceptualising and measuring these core concepts of vulnerability, resilience and adaptive capacity with the aim of translating them into a format suitable for policy implementation. While some contributions seek to fine-tune established theoretical traditions in order to sharpen the conceptualisation of resilience or vulnerability, others seek to synthesise these various streams into a meta-framework that can capture different aspects of adaptation under a singular paradigm (Adger 2006b; Gallopin 2006; Smit and Wandel 2006; Miller et al. 2010; Nelson et al. 2010; Turner II 2010; Hinkel 2011). Across this literature, diverging perspectives draw upon influences from different academic lineages and provide distinct answers to key underlying questions: Who or what is to adapt? How are they to do so? And what are the ends of adaptation? By providing distinct answers to these questions, varied approaches dispute the appropriate sites and scales of adaptation, the rights and responsibilities of affected and contributor groups and the necessary mechanisms and goals of adjustment. Consequently, they legitimate different policy responses and forms of intervention.

At first glance, and despite its coherence at the policy level, such diversity appears to put paid to the notion of climate change adaptation as a discourse. There are sharp debates, for example, over whether core concepts such as resilience are indeed compatible with vulnerability orientated perspectives despite their shared terminologies. This, however, is to miss the forest for the trees. To consider climate change adaptation as a discourse is not to argue that there exists a singular, homogenous and monolithic viewpoint that dominates the field. On the contrary, contrast and contention is intrinsic to any discursive field. Instead, it is to recognise that there is a centre of gravity – a shared set of discursive parameters – around which various perspectives orbit. I argue that the concept of adaptation serves as precisely this centre of gravity. As an overarching concept it constructs the world in terms of a central organising principle predicated on a seemingly natural evolutionary process. So while debate focuses on how adaptation is to be conceived and implemented, the idea of adaptation itself is reproduced as a self-evident and defining point for conceptualising and acting upon social and ecological change in general and climatic change in particular. The discourse precisely constructs and reproduces the shared acceptance of this central organising concept.[2]

A world of adaptation

This centrality of adaptation, I argue, imposes considerable restrictions on the ways in which we can conceive of social transformations in an era of rapid climatic change. Drawn from evolutionary biology and reworked within the parameters of cultural ecology, the concept of adaptation is concerned with the relationship between a species and its environment, both "in the weak sense of articulation with the environment and in the strong sense of evolution to fit the environment" (Harrison 1993: 108). It is this central dichotomy between organism and its environment that grounds the adaptation concept and underscores its purported naturalness. "As environments change", Robert Kates notes, "all life adjusts, adapts and evolves" (2000: 5). When translated into the policy realm, the dualism between 'internal' organism and 'external' environment is retained, yet the notion of society readily substitutes for the idea of an adaptive organism. This switch from species to society enables the framework to link social change to deeply embedded ideas of natural evolution. As cultural ecology frameworks have posited, adaptation is the process by which social systems maintain themselves facing both short-term environmental fluctuations and long-term changes in the composition and structure of their environments (Rappaport 1979: 145).

On this basis, the underlying premise of adaptation is disarmingly simple. In the face of external environmental stimuli such as climatic change, everything and everyone must adapt, materially and culturally, or face unpalatable consequences, from deteriorating life conditions to – potentially – extinction. Undoubtedly within the climate change adaptation literature there is considerable debate over the specific social units that must adapt – i.e. the household, the economy, the community, the region, etc. – and the means of doing so. Some even ask the pertinent questions of who should pay for adaptation, who might be excluded from adaptation, whether adaptation is sustainable or what might lead to 'mal-adaptation'. Yet, there is one commonality that ties this discourse together: adapt we must! For all the diversity within the literature, the idea of adaptation remains unquestioned and forms the conceptual core around which various perspectives orbit. It is from this foundation that the discourse of climate change adaptation constructs a 'world of adaptation' in which every unit, from the household to the nation, can be understood and acted upon in terms of a possessing a latent adaptive capacity to adjust to external environmental shifts levied by anthropogenic climatic change.

Grounded on this axiomatic assumption, other key concepts emerge to populate adaptation's discursive framework. If, by definition, all social actors and ecosystems must adapt to external environmental stimuli, then they must all possess a degree of 'adaptive capacity' as an essential and universal trait. As noted above, adaptive capacity is commonly defined as the ability of a system to adjust to environmental change. However, it is stretched within the climate change adaptation literature to refer more broadly to encompass a variety of social phenomena, including institutions, economies, households and individuals, all of which are seen to possess a relative degree of such capacity to adjust (Hinkel 2011). This provides the concept with a powerful discursive role. As an ontologically derived property of all social actors and institutions, adaptive capacity can subsequently be located, assessed, intervened upon, reshaped and fortified across social space, regardless of historical context or local socio-ecological relations. From the slum inhabitants of Kolkata to the agricultural sector of the Kenyan economy, the idea of adaptive capacity is produced as a new and universal object of development upon which the transformative practices of states, institutions and organisations can be set to work. The task of building adaptive capacity, therefore, provides the universalising glue that binds together climate change adaptation as a core element of governmental practice.

Vulnerability and the dichotomies of adaptation

It is this predisposition to produce a 'world of adaptation' based upon the premise of necessary adjustments to external environmental stimuli that allows the discourse to coalesce and gives it coherence despite its internal tensions. A number of key assumptions, however, are built into the framework of adaptation. First, the adaptation concept constructs its problematic as one of a tension in the relation between two seemingly coherent entities. Climate and society are represented as distinct material realms or systems whose respective stimuli influence and provoke change in each other. Within the paradigm of climate change adaptation, this duality is written into the standard definition of adaptation as "the adjustment in natural or human systems in response to actual or expected climatic stimuli or their effects to moderate harm or exploit beneficial opportunities" (IPCC 2007).

This representation of climate as an external environmental force is most forcibly articulated within the notion of vulnerability, which is the cornerstone concept of the adaptation framework. Omnipresent across the IPCC's summary reports for policymakers (Gaillard 2010), it has given rise to its own diverse and lively academic literature (Adger 2006b: 2010; Miller et al. 2010; Nelson et al. 2010). The centrality of vulnerability stems from the way that it provides a necessary conceptual bridge between the idea of a changing external climate and its impacts upon social dynamics. Habitually defined as the degree to which a 'system' or a particular unit of analysis (household, region, etc.) is susceptible to and unable to cope with adverse effects of climate change (Adger 2006b; IPCC 2007; World Bank 2010), vulnerability provides the conceptual framing that underpins the notion of adaptation as a process of reducing exposure to climatic stress. The relative level of vulnerability that the chosen unit of analysis experiences is commonly broken down into the interaction of three factors: first, the character and magnitude of the exogenous shock or stress; second, the level of sensitivity or defencelessness of the unit; and, third, its adaptive capacity in terms of its ability to moderate potential damages, to take advantage of opportunities or to cope with the consequences of change (IPCC 2001: 995).

While this ubiquitous definition stands as a conceptual centrepiece for the adaptation paradigm, contending perspectives tend to stress one or other side of the equation. Managerial approaches, for example, frequently focus on vulnerability as the potential damage caused by a climatic event, including damage to social and physical infrastructure and potential loss of life (United Nations Framework Convention on Climate Change 2007; World Bank 2008). This framing emphasises vulnerability as a state conditioned by the character and frequency of a given physical hazard and the degree of exposure and sensitivity of an associated social or ecological system. It therein gravitates towards resolutely technocratic solutions, such as reinforcing flood defences or promoting the substitution of current crops with drought-resistant varieties (e.g. World

Bank 2006b). In contrast, more critically orientated perspectives emphasise vulnerability as a socially constructed defencelessness that exists prior to shocks (O'Brien et al. 2007; Ensor and Berger 2009). This latter approach usefully highlights issues of marginalisation and inequality that unevenly stratify the impacts of climate-related shocks between social groups. Such 'contextual vulnerability' is seen to translate directly into unequal degrees of exposure to the climate hazard and produces strongly differentiated degrees of adaptive capacity with which to deal with it (Brooks 2003).

In some respects, these two approaches seem worlds apart and important policy differences duly emerge from their analytical cleavages (see Chapter Four). Notwithstanding such divergences, however, there remains a central commonality wherein both framings of vulnerability remain predicated on an internal–external dichotomy between society and climate. As visualised in standardised vulnerability models (see Figure 3.1), this constructs climate change as a coherent external force that delivers external environmental shocks to an otherwise independent social system. The impacts of these shocks are then mediated by the particular character of local social structures and institutions (see Brklacich, Chazan and Bohle 2010). As Ben Wisner and others clearly put it in their seminal account of risk, disasters emerge at the intersection of two opposing forces: "those processes generating vulnerability on one side, and the natural hazard event (or sometimes a slowly unfolding natural process) on the other" (Wisner et al. 2004: 50). The

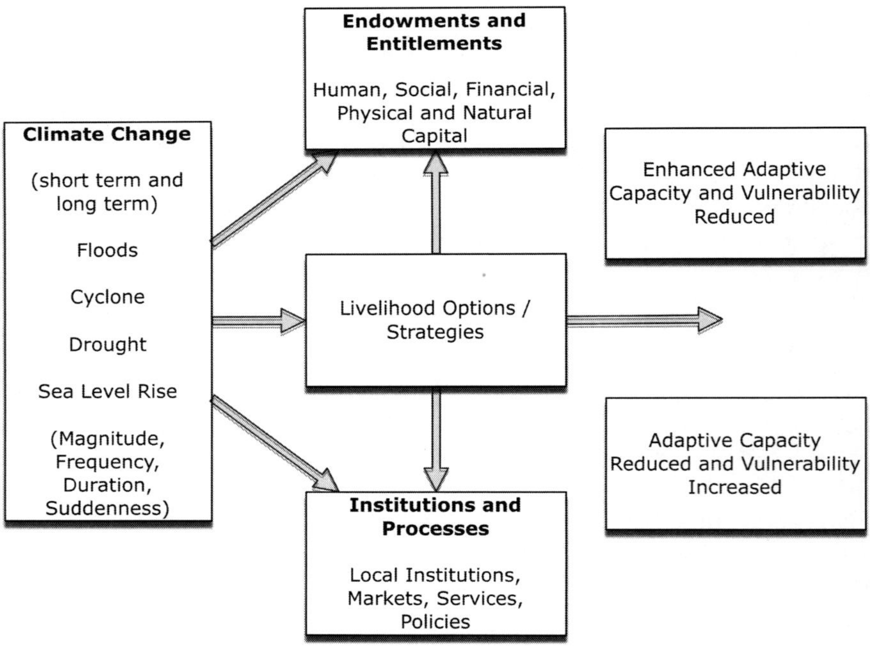

Figure 3.1 Model of vulnerability, drawn from Dulal et al. (2010)

two are seen to come together – like a nutcracker – to cause substantial yet uneven social disruption and physical damage.[3]

As plausible as it might seem at first glance, once we step back from the representation of climate change as a coherent external disturbance to an otherwise independent society, the foundational distinctions upon which the analytical edifice is built become strained. To render climatic change as an external threat to an otherwise coherent society requires bracketing off and abstracting climate as something that is outside of society as part of its exterior environment. The dichotomy between natural hazard and social exposure obscures the complex processes of co-production in which biophysical forces and social energies are intimately and indivisibly interweaved in the creation of lived environments (cf. Smith 1984; Latour 1993; Castree 2001; Swyngedouw 2004). As an example, consider the tragic floods of 2013 that afflicted the northern Indian state of Uttarakhand, killing thousands of people and severely damaging physical infrastructure, urban settlements and agricultural lands. In mid-June of that year, intense precipitation from monsoon storms engorged mountain streams while, in the context of above-average seasonal temperatures, melting glaciers released further flows. Together, these processes led to steeply rising rivers both upstream and downstream and to rockslides in the high ranges that temporarily created pools only to later release cascades of pent-up waters. The violent flows of water that coursed downstream resulted in considerable destruction, not least when they broke existing riverbanks and reverted to former paths, obliterating the built environments that now colonised those reclaimed lands (Parkash 2013).

In the standard adaptation model noted above, the region could be described as experiencing a strong external shock to its social and natural systems. Both the latter, in turn, displayed considerable vulnerability to this external stress owing to high sensitivity and limited adaptive capacity. The resulting outcome was a disaster that entailed considerable devastation to infrastructure, ecosystems and livelihoods, including a tragic loss of human lives. That the intensity of the rainfall appears to be related to changing climatic patterns seemingly supports the instantiation that the Uttarakhand disaster exemplifies a latent vulnerability to climate change characteristic of the Himalayan region at large (Balasubramanian and Kumar 2014). In response, a process of planned adaptation to such external threats emerges as the logical outcome, a process that requires concerted governmental actions such as building flood defences and relocating vulnerable settlements (Government of Uttarakhand 2012). More critically orientated perspectives would also emphasise that social inequalities within the Himalaya unevenly shape the relative vulnerability of distinct social groups, therefore necessitating targeted adaptation policies that can address the specific levels of exposure to climatic threats faced by distinct sections of the population.

At closer inspection, however, the constitution of this event is more complex than such a framework can convey. The excess of water at the heart of the disaster – the supposed 'natural hazard event' – can only be considered excessive or hazardous in relation to the socio-ecology of the landscapes it encountered in its flows. In this respect, changing rainfall patterns were simply one aspect of an ongoing socio-ecological transformation that, cumulatively, produced the flood. Over the previous four decades, a plethora of small and large dams built for hydroelectric generation had radically re-orientated the hydrology of the region (Agarwal and Narain 1987; Agarwal 2013). Although they temporarily held up water flows, once breeched such dams sent surges of water downstream that carried large amounts of construction silt, boulders and other debris that acted like sandpaper to exponentially increase downstream erosion (Parkash 2013). It was this material acting in conjunction with the water that wreaked destruction upon the houses that crowded the riverbanks and obliterated cultivated fields. At the same time, significant deforestation in upland areas for both hydropower projects and extensive mining had changed the absorption capacity of hillside soils while removing barriers to rain runoff. Massive road construction to promote tourism further weakened hillsides, making them more prone to landslides, while simultaneously fuelling the expansion of urbanised areas downstream that lay in the flood path (Balasubramanian and Kumar 2014). Such urbanisation was conspicuously manifested in the construction of hotels and tourist-orientated lodging directly upon the scenic riverbanks and reclaimed flood plains (Agarwal 2013).

In short, there was nothing essentially 'climatic' or 'natural' about how heavy rainfall translated into what the adaptation discourse might term an 'external shock'. It was not simply a case of a climatically generated 'natural hazard' combining with 'social defenceless' to cause a disaster because the very hazardness of the deluge can only be conceived in relation to the produced environment and its underlying socio-ecology. This indicates the need to go beyond the simplicity of external shock metaphors to conceptualise how varied biophysical processes – including those labelled as 'climatic' – are inscribed within an environment as part of a continuous and dynamic process of socio-ecological production. To do so greatly complicates the adaptation model. On the one hand, it questions how we can view meteorological processes as external to a historically formed lived environment within which they are essential productive components. Are the drops of rain that fall on the upslopes of the Himalayas 'external', 'natural' or 'climatic' up until the moment they touch the ground and run into the manufactured regional watershed, where they suddenly become socialised as they are fed into agriculture, hydropower and tourism? On the other, it questions how we can pre-assign causality to such biophysical drivers as a 'natural hazard', rendering other socio-ecological forces as simply contextual factors. The power of rainfall to act as a hazard is inherently predicated upon a produced socio-ecological context that precedes it. The very concept of 'natural hazard' implies an innately socialised nature.

Instead of an underlying condition of social vulnerability to climate change, what the tragic Uttarakhand experience demonstrates is the degree to which the Himalayan region is an intensely produced environment that confounds any clear boundaries between 'internal' and 'external', 'social' and 'climatic'. The recent transformations that the region has experienced – habitually termed 'development' or 'modernisation' – are predicated upon an amalgam of social energies and biophysical forces operating across spatial scales that had been closely worked into the active production of the landscape as a lived environment. Mining and river dredging, for example, serve to claim stones, sand and other materials carved by erosion that subsequently facilitate the local and distant production of built environments. Agro-forestry is predicated upon harnessing prevailing geo-climatic conditions to buttress a tense accommodation between local communities and powerful commercial interests. Wide-scale road construction facilitates the inflows of tourists who derive religious meaning from the hydrological sources of the Ganges and whose growing presence has unevenly reconfigured livelihoods across the region. The plethora of dams and turbines, moreover, determinedly channel water to produce electricity that haltingly powers the ongoing transformation of India under rapid urbanisation and staggered industrialisation.

The experience of climatic change, therefore, emerges not as a coherent external force or shock. Rather, it manifests itself through these dynamic processes of socio-ecological production. It is bound up within contrasting forms of agency that cut across geographic scales and nature-social binaries. Household livelihood strategies, circuits of capital accumulation and the exercise of state power are manifested in and through water flows, erosion, glacial melt and the biotic and abiotic processes that underpin agriculture, forestry and power generation. From this perspective, the rubric of external climatic shocks or disturbances does scant justice to the mutual entanglements of biophysical processes and social organisation that render the lived environment productive for particular human purposes while, concurrently, forging a landscape that can prove brutally destructive to those same ends (see also, Mustafa 2005; Collins 2010). To maintain the framework of external relations between society and nature that underpin the adaptation discourse, humans must be seen as having 'adapted' the Himalayan environment – including its climatic dimensions – for particular purposes and then must equally 'adapt' themselves to the new configuration of socio-ecological processes ending in an ongoing process of mutual 'co-adaptation'. The troubled discursive contortions necessary to maintain this framework of external influences indicate deep tensions residing at the heart of the adaptation paradigm. As Raymond Williams once noted, "[w]e have mixed our labour with the earth, our forces with its forces too deeply to be able to draw back and separate either out" (1980: 83).

Unmaking a world of adaptation

The danger of representing the valleys of Uttarakhand in terms of vulnerability to an exogenous climatic threat is clear in its tendency to naturalise the ongoing production of the lived environment while exceptionalising climate as a coherent external force. In this respect, the discourse of adaptation is intimately bound up with the representational politics that prefigure institutional actions. To represent climatic change as an external disturbance is an inherently political act of boundary making that separates out 'external climate' from 'internal society' in a way that rationalises specific forms of governance and social transformation and simultaneously obscures others. Out of this discursive production of climate as an externality there emerge a series of seemingly natural boundaries – societies, communities, ecosystems, economic sectors – that face an exogenous stress to which they must adapt. With climate posited as an external threat to a pre-existing state of normality, then the politics of building adaptive capacity to lessen vulnerability and promote 'resilience' or 'human security' surges forward as a similarly 'natural' response. This is why the discourse of adaptation is particularly germane to a technocratic politics of intervention that maintains what it denotes as the 'natural' parameters of the lived environment.

As the Uttarakhand example showed, however, what appears as nature or natural is already shaped by forms of power, technology, expertise and privilege (Mitchell 2002: 210; see also, Castree 2014). By approaching climatic change from a perspective focused on socio-ecological production, it is possible to disrupt the internal-external dichotomy used by the adaptation discourse to create a clear notion of climatic causality to which society must then adapt. It indicates the pressing need to reject the externalisation of climate and instead understand how meteorological forces are indelibly written into the active production and transformation of a region as a lived, yet inherently unequal, environment. In place of an imagery of a changing external environment to which society must then adapt, we face landscapes that are simultaneously social and ecological and that are brought into existence through processes of continual production and transformation within which meteorological forces are deeply complicit. Different social actors are incorporated into these active processes of environmental production in varied and extremely uneven ways, meaning that any process of planned transformation will cut across conflicting interests, raising distinct political challenges. In Uttarakhand, this is manifest in a raft of struggles over forest management, the damming and control of rivers, the extent of mining and agricultural labour that spans divisions of class, caste and gender in complex ways (Pathak 1997; Guha 2000; Rangan 2000; Gururani 2002; Agrawal 2005; Linkenbach 2007; Agarwal 2013).

These concerns form the heart of what I term a 'lived environment' approach, as set out in Chapter One. To start from this perspective is to create a methodological approach that is markedly distinct from that of the adaptation framework. There is no static division between humans and their environments on which any clear-cut process of adaptation can be said to occur. Instead, we can only reach an adequate understanding of climatic change by first asking how diverse meteorological processes are worked into the production of lived environments as part of a complex of social and biophysical forces that operate on multiple, overlapping scales. To do so requires a resolutely historical approach in order to addresses pertinent questions of how power and scale infuse the production of lived environments. From this departure point a different set of core questions arise. What are the specific entanglements of meteorological forces and social processes that produce the lived environment? What spatial scales and temporal horizons do these processes operate upon? And what social cleavages and forms of power are built into and reproduced within the resulting lived environments?

The discourse of climate change adaptation, however, is configured to do the opposite. It tends to bring the socio-ecological processes of production to a shuddering halt by freeze-framing the ongoing transformation of lived environments so that it can isolate and extract 'climate change' from this snapshot as a causal agent that possesses its own dynamics in separation from the socio-ecological integuments that make it tangible. Once suitably shorn of such context, climate is then reinserted as a series of projected external shocks and disturbances to which social agents can then be judged in terms of their vulnerability or resilience and to which their adaptive capacity can be fortified. Subsequently, the socio-ecological context that was extracted at the beginning can then be brought back in at the end, packaged in terms of the degree of vulnerability or resilience to external climatic stimuli that a given unit of analysis possesses. Through this representational regime, adaptation is produced as a field of governmental action open to a specific form of social engineering aimed at reducing vulnerabilities and building resilience in the face of supposedly exogenous threats.

If, however, we resist the urge to press pause on the active transformation of lived environments, the concept of adaptation stutters. The idea that society must adapt to climatic change becomes banal: a truism that reveals very little and yet already assumes too much. It is only by discursively externalising climate that the latter can be represented as a coherent and external force that provides stimuli and shocks to an otherwise distinct social system. Once we loosen the conceptual boundaries inherent to this discursive framework, climate becomes irreducible to a realm of external influences. As the Uttarakhand example highlighted, climate change needs to be understood in terms of how meteorological forces are intrinsically interwoven into the continual transformative activities that produce lived environments in both their social and climatic dimensions. It is in this context of active socio-ecological transformation that we can understand fully the ways in which climatic forces are rendered simultaneous creative and destructive. The power accorded to meteorological phenomena derives not from their 'natural' properties but from the way they are situated within the field of socio-ecological relations of which they form an irreducible part (Moore 2013).

What we term 'anthropogenic climate change' needs to be considered in this context. It represents one further moment in the ceaseless environmental reconfiguration that is historically embedded within the socio-ecology of capitalist modernity to a degree and pace that is both unparalleled and accelerating. As a consequence, when Leary et al. (2010: 8) suggest that "current practices, processes, systems and infrastructure that are more or less adapted to the present climate will become increasingly inappropriate and maladapted as the climate changes," this appears to overlook the immense processes of socio-ecological transformation that are ongoing across social space and which render notions of an 'adapted present' specious. Both agrarian and urban environments find themselves in constant flux, driven by the undulating rhythms of capital accumulation, technological change, the flows of commodities and human bodies and contested political practices that continually reconfigure social space. Climatic change is not an exception or externality to these processes. It makes no sense to detach it as some sort of independent variable. On the contrary, it is a further enduring element – both cause and effect – of the active production of humans as part of obdurately uneven lived environments. Indeed, given the impossibility of neatly separating out these processes, our attention is better served by asking why the discourse of adaptation rests on and reproduces a conceptual framework that orders socio-ecological relations in this dichotomous fashion.

Adaptation, governmentality and power

The answer, I believe, can be found in the enduring institutional need to fashion climate change as a distinct realm of governmentality in which processes of change can seemingly be circumscribed, managed and controlled. As tersely put by Erik Swyngedouw, the institutional response to climate change centres upon a techno-managerial apparatus that combines emergent eco-technologies with institutional configurations that collectively seek "a socio-ecological fix to make sure nothing really changes" (2010: 222). As noted in the examination of discourses and practices of adaptation in the context of Pakistan, India and Mongolia in Chapters Six through Eight, the motif of controlled change to make sure things stay the same is deeply embedded within the institutionalisation of adaptation. In representing the issue in terms of a fundamental conflict between society and nature, the paradigm of adaptation is well suited to these purposes. This is because, as Raymond Williams argued, such an approach "spares us the effort of looking, in any active way, at the whole complex of social and natural relationships which is at once our product and [the preconditions for] our activity" (1980: 283).

By abstracting from the ways in which humans co-produce their lived environments across scales, the concepts of vulnerability, resilience and adaptive capacity overwhelmingly are portrayed as issues of 'here and now' in which questions of adaptation are radically separated from trajectories of socio-ecological change that have a longer time frame and whose causative forces stretch beyond the places in which they manifest themselves. The various incarnations of the IPCC reports, starting in the mid-1990s up to 2007, for example, dutifully recognise the importance of "non-climatic forces and conditions" that impact upon the constitution of vulnerability and shape any adaptation process. Yet these dynamics are typically reduced, in the words of Bassett and Fogelman (2013), to "proximate factors whose causal roots are never theorized". As such, the troubled history of how our present world has been brought into being, with its deep historical fissures and unequal power over the use and consumption of resources, is naturalised alongside the simultaneous exceptionalisation of climatic change. On this basis, as Elizabeth Shove puts it, "policy proceeds on the basis of a characteristically thin account of the social world" (2010: 277).

Such thin accounts of the social world are necessary to render climatic change governable. It is by extracting the 'here and now' of climatic change from the densely layered dynamics of historical processes that the discourse produces the concepts of vulnerability, adaptive capacity and resilience as 'travelling rationalities' that seemingly possess universal applicability as technical categories that are applicable regardless of historically formed local context (Mosse 2011). 'Adaptive capacity' and 'vulnerability', as noted above, are represented as inherent internal properties of any unit of analysis that must adapt to external stimuli. As categories presumed to be ontologically grounded features of all societies threatened by external climatic change, the adaptation paradigm maintains the idea of climate change adaptation as a local example of a process that is general, universal and infinitely scalable. At varied points in the literature – and often within the same article – the unit of vulnerability or adaptation can be the individual, the household, the community, the social group, a city, an economic sector, a region or the nation (see for example, the skipping between scales in Smit and Wandel 2006; Miller et al. 2010). Little attention seems to be given to the fact that the meanings and measures of vulnerability and adaptive capacity would need to change dramatically from one unit to the other and from one context to another. To talk of a household or a region or an economic sector being 'vulnerable' is to conceptualise qualitatively distinct processes that are expressive of very different socio-ecological relations. As such, the term vulnerability frequently operates as a fungible placeholder that reinforces the idea of adaptation as universal process that simply requires a localised fine-tuning. The cumulative force of these tendencies is to create a marginalisation of questions of power within the discourse, a trend that steadfastly opens the political terrain for technocratic colonisation.

The exteriorisation of knowledge production

The discourse of adaptation, therefore, responds closely to an institutional impetus to produce narratives of climate change that both profess universal validity and establish clear boundaries upon which to stabilise policy recommendations. Such an approach facilitates the exteriorisation of knowledge production in a manner that radically separates the creation of scientific knowledge on climatic change from the institutional practices and political designs of the agencies involved. The World Bank, for example, is now a keen producer of knowledge and strategies on climate change adaptation. Notably, however, the bank is concurrently an actor that has steadfastly sought to reshape directly the global political economy through programmes of economic liberalisation, state transformation, natural resource mobilisation and biotechnology promotion and indirectly through the funding of carbon-powered industrialisation projects, not least in the form of coal-fired power stations and the propagation of ineffective carbon markets (see Bumpus and Liverman 2010; Zacune 2011). In the context of agrarian regions with which much of this book is focused, the World Bank has a very specific and contested vision of rural transformations predicated upon managing the transition out of agriculture for rural populations whose labour is deemed as surplus to the requirements of a suitably rationalised and entrepreneurial agricultural sector (World Bank 2007: see Chapter Five). These overlapping roles, the tensions and contradictions between them and their impacts upon the framing of 'scientific knowledge' are simply not acknowledged by the institution.

As Mitchell argued in a different but related context, international development has "a special need to overlook this internal involvement in the places and problems it analyzes, and present itself instead as an external intelligence that stands outside the objects it describes" (2002: 211). In this respect, the World Bank embodies precisely the kinds of contradictions that the discourse of climate change adaptation occludes: namely, a disposition to avoid situating climatic change within a broad and suitably historicised understanding of societal reproduction at interlinking scales of analysis. In part, such closures reflect the tendency of policy-orientated analysis to seek a clearly delineated 'object of development' in which cause-and-effect can be easily demarcated and interventions with predictable results planned. In the professionalised world of development practice, as Piers Blaikie put it, there is a paramount need for narratives that are "fairly simple, elegant, and appealingly told so as to resonate with the professional and cultural repertoires of their constituencies" (2000: 1041).

The intrinsic 'thinness' of climatic change that such accounts produce, however, reflects not only prevailing institutional logics for simplified managerial frameworks but also the power relations within, between and across states through which knowledge is produced, legitimated and dispersed. Considerable energies are placed into the production and generalisation of climate change adaptation as a universal and depoliticising representational regime precisely because of the heavily politicised nature of the issues at hand (see Swyngedouw 2010; Castree and Felli 2012). A number of social theorists, for example, have argued that climate change has the potential to interrupt embedded understandings of society, nature and the social production of risk (Beck 2010; Shove 2010). As Chakrabarty puts it, in becoming an agent of macro-climatic transformation, industrial capitalism has acted like the rabbit hole in Alice's story:

> [W]e have slid into a state of things that forces on us a recognition of some of the parametric (that is, boundary) conditions for the existence of institutions central to our idea of modernity and the meanings we derive from them.
> (Chakrabarty 2009: 52)

The potential ruptures, however, are greater than Chakrabarty's account acknowledges. The spectre of anthropogenic climate change opens a window onto a decidedly stratified global socio-ecology in which the production of resources, their consumption, the distribution of their waste products and the ensuing gains and risks involved in such processes are brutally uneven in their distribution within and across nation-states (Peet, Robbins and Watts 2011). By foregrounding the uneven cartographies of production and consumption, the spectre of climatic change serves to raise contrasting and often irreconcilable normative ideas of value, sustainability, development, ownership, security and control. In short, climate change raises dirty questions concerning both the future of capitalism as a form of organising socio-ecological relations on a global scale while bringing new attention onto the uneven legacies of its historical past (Clark and York 2005; Moore 2011). As Nigel Clark has noted, this is not a question of a rupture between society and nature but of a produced socio-ecological metabolism that is both global in scope and that is predicated upon an "overly effective" tapping and channelling of energies that entail fearsome consequences far beyond the intentions of any specific participants (2010: 45).

Where for art thou capitalism?

In this respect, the spectre of climatic change could be said to haunt contemporary capitalism in a manner quite different, yet perhaps no less vitally, to the one that Marx saw haunting the European bourgeoisie in the mid-nineteenth century. It is, therefore, conspicuous how the discourse of climate change adaptation studiously avoids bringing the relationship between capitalism and climatic change into focus. Tellingly, the portrayal of climate as external force creates the ontological grounds on which to separate the outcomes of climatic change from the anthropogenic dimensions of its production. It swiftly binds the discussion of the production of climatic change under the field of 'mitigation' and the discussion of its impacts under the rubric of 'adaptation'. The two are occasionally brought back together through discussion of potential synergies between adaptation and mitigation initiatives. Yet this is an impoverished conceptualisation in which the happenstance occasion of 'win-win' overlaps obscures any deeper consideration of climatic change as integral to the production of climate under contemporary capitalism. Not least, in constructing a 'world of adaptation', the politics of adaptation can be tightly separated from considerations of the production and consumption of commodities and their associated metabolic processes and how those are articulated within changing forms of capital accumulation on local and more expansive scales. It spectacularly avoids consideration of how climatic change is intimately interwoven with the construction of a global apparatus of market rule through which the human and natural resources necessary for expanded capital accumulation are incorporated through the power of institutionally backed financial claims into strikingly uneven global divisions of production and consumption (Taylor 2009).

It, therefore, is profoundly concerning how, within the adaptation frame, the inherently unequal socio-ecological transformations that are innate to the dynamics of capitalism are reduced to static contextual factors that provide a tapestry upon which climate change sets to work and adaptation may precede. While the discourse of adaptation has little place for it, climatic change must surely be understood in the historical context of ongoing systematic transformation of lived environments under the weight of the troubling socio-ecological dynamics that rivet the era of global capitalism. This is not to renege into a black-and-white analysis that simply passes the cap of responsibility onto an abstract force named capitalism. Rather it requires a resolutely historical approach to tease out the pertinent socio-ecological energies and forms of agency at work in reshaping our world. Without so doing, we face a problematic myopia that obscures how questions of 'adaptation' – i.e. the power to reshape socio-ecological relations – are intimately intertwined with the production of climatic change and our tenuous abilities to reconfigure our own social metabolisms in its wake. This is precisely the troubled and troubling terrain in which the discourse of climate change adaptation has emerged. It is also one that the discourse studiously marginalises. As a consequence, technocratic and managerial responses find it easy to occupy its terrain no matter how diligently some critical adaptation theorists seek to contest it. The following chapter elaborates this point in further detail.

Notes

1 I examine the concept of resilience in the following chapter.
2 There is parallel here to Escobar's discussion of development as a discourse. For Escobar, the discourse of development was centred upon the question of poverty which acted as a homogenising organising concept under which the diverse histories, social trajectories and ways of living of peoples across the postcolonial world (and further afield) could be represented and then tied to its associated meta-narrative of 'development'. As such, the problematisation of poverty conjured development as a universal cure to its abnormalities: "That the essential trait of the Third World was its poverty and that the solution was economic growth and development became self-evident, necessary, and universal truths" (Escobar 1995: 24). In the discursive structure of climate change adaptation, vulnerability plays a broadly comparable role to that of poverty and adaptation approximates development. Escobar's problem was that he tended to fetishise the discursive parameters he identified. Rather than see the discourse of development as dynamic and contested, with periods of both stability and change, he instead rigidly bound discursive structures, institutional actors and subject formation together into an unflinching discourse that appeared curiously impervious to contradiction, struggle and change. What was lost in Escobar's original postulation of the discourse of development, therefore, was that discursive structure can only imperfectly and haltingly have the effects he accorded to it (see Gidwani 2002: for a superb critique).
3 It should be noted that Blaikie et al.'s original framework positioned the social dimensions of such vulnerability within encompassing 'structures of domination' that emphasised both the relational dynamics of vulnerability alongside its multi-scalar dimensions. These aspects are often lacking in the more recent climate change-orientated discussions of the social dimensions of vulnerability, which tend to focus on static inequalities, as discussed in chapter four.

References

Adger, W. N. (Ed.). 2006a. *Fairness in Adaptation to Climate Change*. Cambridge, MA: MIT Press.
Adger, W. N. 2006b. "Vulnerability." *Global Environmental Change* 16(3): 268–81.
Adger, W. N., I. Lorenzoni, and K. O'Brien. 2010. "Adaptation Now." Pp. 1–22 in *Adapting to Climate Change*, edited by W. Neil Adger, I. Lorenzoni, and K. O'Brien. Cambridge: Cambridge University Press.
Agrawal, A. 2005. *Environmentality: Technologies of Government and the Making of Subjects*. Durham, NC: Duke University Press.
Agarwal, A., and S. Narain. 1987. *Floods, Flood Plains and Environmental Myths*. New Delhi: Centre for Science and Environment.
Agrawal, A., and N. Perrin. 2008. "Climate Adaptation, Local Institutions, and Rural Livelihoods." *IFRI Working Paper* (# W08I-6): 1–17.
Agarwal, R. 2013. "Hydropower projects in Uttarakhand: Displacing people and destroying lives." *Economic and Political Weekly* XLVIII(29): 14–18.

Balasubramanian, M., and P.J.D. Kumar. 2014. "Climate change, Uttarakhand and the World Bank's message." *Economic and Political Weekly* XLIX(1): 65–69.

Bassett, T., and C. Fogelman. 2013. "Deja vu or something new? The adaptation concept in the climate change literature." *Geoforum* 43(1): 42–54.

Beck, U. 2010. "Climate for Change, or how to create a green modernity?" *Theory, Culture and Society* 27(2/3): 254–66.

Bizikova, L., S. Burch, S. Cohen, and J. Robinson. 2010. "Linking Sustainable Development with Climate Change Adaptation and Mitigation." Pp. 157–79 in *Climate Change, Ethics and Human Security*, edited by Karen O'Brien, Asuncion Lera St. Clair, and Berit Kristoffersen. Cambridge: Cambridge University Press.

Blaikie, P. 2000. "Development, post-, anti-, and populist: A critical review." *Environment and Planning* A 32(4): 1033–50.

Boyd, E., N. Grist, S. Juhola, and V. Nelson. 2009. "Exploring development futures in a changing climate: Frontiers for development policy and practice." *Development Policy Review* 27(6): 659–74.

Brklacich, M., M. Chazan, and H. Bohle. 2010. "Human Security, Vulnerability and Global Environmental Change." Pp. 35–51 in *Global Environmental Change and Human Security*, edited by R. Matthew, J. Barnett, B. McDonald, and K. O'Brien. Cambridge, MA: MIT Press.

Brooks, N. 2003. "Vulnerability, risk and adaptation: A conceptual framework." *Tyndall Centre for Climate Change Research, Working Paper* 1(38): 1–20.

Bumpus, A. G., and D. M. Liverman. 2010. "Carbon Colonialism? Offsets, Greenhouse Gas Reductions and Sustainable Development." Pp. 203–27 in *Global Political Ecology*, edited by Michael Watts, Richard Peet, and Paul Robbins. London: Routledge.

Burton, I., L. Bizikova, T. Dickinson, and Y. Howard. 2007. "Integrating adaptation into policy: Upscaling evidence from local to global." *Climate Policy* 7(4): 371–76.

Castree, N. 2001. "Socializing Nature: Theory, Practice and Politics." Pp. 1–21 in *Social Nature: Theory, Practice and Politics*, edited by Noel Castree and Bruce Braun. Oxford: Blackwell.

Castree, N. 2014. *Making sense of nature: representation, politics and democracy*. London: Routledge.

Castree, N., and R. Felli. 2012. "Neoliberalising Adaptation to Environmental Change: Foresight or Foreclosure." *Environment and Planning* A 44(1): 1–4.

Chakrabarty, D. 2009. "The climate of history: Four theses." *Critical Inquiry* 35(4): 197–226.

Chakrabarty, D. 2012. "Postcolonial studies and the challenge of climate change." *New Literary History* 43(1): 1–18.

Clark, B., and R. York. 2005. "Carbon metabolism: Global capitalism, climate change, and the biospheric rift." *Theory and Society* 34(4): 391–428.

Clark, N. 2010. "Volatile worlds, vulnerable bodies: Confronting abrupt climate change." *Theory, Culture and Society* 27(2–3): 31–53.

Collins, T. 2010. "Marginalization, facilitation and the production of unequal risk: The 2006 Paso del Norte Floods." *Antipode* 42(2): 258–88.

Dulal, H., G. Brodnig, C. Onoriose, and H. Thakur. 2010. "Capitalizing on assets: Vulnerability and adaptation to climate change in Nepal." *World Bank Social Development Papers: Social Dimensions of Climate Change* January 2010(121): 1–23.

Ensor, J., and R. Berger. 2009. *Understanding Climate Change Adaptation: Lessons from Community-Based Approaches*. Rugby, UK: Practical Action Publishing.

Escobar, A. 1995. "Encountering Development: The Making and Unmaking of the Third World." Princeton, NJ: Princeton University Press.

Fisher, G., M. Shah, F. Tubiello, and H. van Velhuizen. 2005. "Socio-economic and climate change impacts on agriculture: an integrated assessment, 1990–2080." *Philosophical Transactions of the Royal Society* 360(1463): 2067–83.

Gaillard, J. C. 2010. "Vulnerability, capacity and resilience: Perspectives for climate and development policy." *Journal of International Development* 22(2): 218–32.

Gallopin, G. 2006. "Linkages between vulnerability, resilience, and adaptive capacity." *Global Environmental Change* 16(2): 293–303.

Gidwani, V. 2002. "New theory or new dogma? A tale of social capital and economic development from Gujarat, India." *Journal of Asian and African Studies* 37(2): 83–112.

Government of Nepal. 2011. "Climate Resilient Planning: A Tool for Long Term Climate Adaptation." Edited by National Planning Commission. Kathmandu: Government of Nepal.

Government of Uttarakhand. 2012. *State Action Plan on Climate Change: Transforming Crisis into Opportunity*. Dehradun: Forest Department Uttarakhand.

Guha, R. 2000. *The Unquiet Woods: Ecological Change and Peasant Resistance in the Himalaya*. Berkeley: University of California Press.

Gururani, S. 2002. "Forests of pleasure and pain: Gendered practices of labor and livelihood in the forests of the Kumaon Himalayas, India." *Gender, Place & Culture* 9(3): 229–43.

Harrison, G. A. 1993. *Human adaptation*. Oxford: Oxford University Press.

Hinkel, J. 2011. "Indicators of vulnerability and adaptive capacity: Towards a clarification of the science-policy interface." *Global Environmental Change* 21: 198–208.

Huq, S., and H. Reid. 2004. "Mainstreaming adaptation in development." *IDS Bulletin* 35(3): 15–21.

IPCC. 2001. *Climate Change 2001: Impacts, Adaptation and Vulnerability. Summary for Policy Makers*. Geneva: World Meteorological Organisation.

IPCC. 2007. *Climate Change 2007: The Scientific Basis. WGI Contribution to IPCC 4th Assessment Report*. Cambridge: Cambridge University Press.

Ireland, P. 2012. "Climate change adaptation: Business-as-usual aid and development or an emerging discourse for change?" *International Journal of Development Issues* 11(2): 92–110.

Kates, R. 2000. "Cautionary tales: Adaptation and the global poor." *Climatic Change* 45(1): 5–17.

Latour, B. 1993. *We Have Never Been Modern*. Cambridge, MA: Harvard University Press.

Leary, N., J. Adejuwon, V. Barros, P. Batima, B. Biagini, I. Burton, . . . M. Wehbe. 2010. "A Stitch in Time: General Lessons from Specific Cases." Pp. 1–27 in *Climate Change and Adaptation*, edited by Neil Leary et al. London: Earthscan Press.

Linkenbach, A. 2007. *Forest Futures: Global Representations and Ground Realities in the Himalayas*. London: Palgrave.

Miller, F., H. Osbahr, E. Boyd, F. Thomalla, S. Bharwani, G. Ziervogel, B. Walker, J. Birkmann, S. van der Leeuw, J. Rockstrom, J. Hinkel, T. Downing, C. Folke, and D. Nelson. 2010. "Resilience and vulnerability: Complementary of conflicting concepts." *Ecology and Society* 15(3): 11–37.

Ministry of Environment and Forests Bangladesh. 2008. *Bangladesh Climate Change Strategy and Action Plan 2008*. Dhaka: Government of the People's Republic of Bangladesh.

Mitchell, T. 2002. *Rule of experts: Egypt, Techno-Politics, Modernity*. Berkeley: University of California Press.

Moore, J. 2011. "Transcending the metabolic rift: a theory of crises in the capitalist world-ecology." *Journal of Peasant Studies* 38(1): 1–46.

Moore, J. 2013. "From Object to Oikeios: Environment-Making in the Capitalist World-Ecology." Unpublished paper. Department of Sociology, Binghamton University. Available at: http://www.jasonwmoore.com/uploads/Moore__From_Object_to_Oikeios__for_website__May_2013.pdf accessed October 1st, 2013.

Morton, J. 2007. "The impact of climate change on smallholder and subsistence agriculture." *Proceedings of the National Academy of Sciences of the United States of America* 104(50): 19680–85.

Mosse, D. 2011. "The Anthropology of Expertise and Professionals in International Development." Pp. 1–31 in *Adventures in Aidland: The Anthropology of Professionals in International Development*, edited by David Mosse. New York: Berghahn Books.

Mustafa, D. 2005. "The production of an urban hazardscape in Pakistan: Modernity, vulnerability, and the range of choice." *Annals of the Association of American Geographers* 95(3): 566–86.

Nelson, R., P. Kokic, S. Crimp, H. Meinke, and S. M. Howden. 2010. "The vulnerability of Australian rural communities to climate variability and change: Part I – conceptualising and measuring vulnerability." *Environmental Science and Policy* 13(1): 8–17.

O'Brien, K., S. Eriksen, L. Nygaard, and A. Schojolden. 2007. "Why different interpretations of vulnerability matter in climate change discourses." *Climate Policy* 7(1): 73–88.

OECD. 2009. "Policy Guidance on Integrating Climate Change Adaptation into Development Co-operation." Edited by Development Assistance Committee. Geneva: OECD.

Orlove, B. 2009. "The Past, the Present and Some Possible Futures of Adaptation." Pp. 131–63 in *Adapting to Climate Change: Thresholds, Values, Governance*, edited by W. Neil Adger, Irene Lorenzoni, and Karen O'Brien. Cambridge: Cambridge University Press.

Parkash, S. 2013. "Brief Report on Uttarakhand Disaster (16/17 June 2013)." Edited by National Institute of Disaster Management. Delhi: National Institute of Disaster Management.

Pathak, S. 1997. "State, society and natural resources in the Himalaya: Dynamics of change in colonial and post-colonial Uttarakhand." *Economic and Political Weekly* 32(17): 908–12.

Peet, R., P. Robbins, and M. Watts. 2011. *Global Political Ecology*. Abingdon, Oxon; New York: Routledge.

Pelling, M. 2011. *Adaptation to Climate Change*. London: Routledge Press.

Rangan, H. 2000. "State Economic Policies and Changing Regional Landscapes in the Uttarakhand Himalaya." Pp. 23–46 in *Agrarian Environments: Resources, Representation and Rule in India*, edited by Arun Agrawal and K. Sivaramakrishnan. Durham, NC: Duke University Press.

Rappaport, R. A. 1979. *Ecology, Meaning, and Religion*. Richmond, CA: North Atlantic Books.

Shove, E. 2010. "Social theory and climate change: Questions often, sometimes and not yet asked." *Theory, Culture and Society* 27(2–3): 277–88.

Smit, B., and J. Wandel. 2006. "Adaptation, adaptive capacity and vulnerability." *Global Environmental Change* 16(3): 282–92.

Smith, N. 1984. *Uneven Development: Nature, Capital, and the Production of Space*. New York: Blackwell.

Stern, N. 2007. *The Economics of Climate Change: The Stern Review*. Cambridge: Cambridge University Press.

Swart, R., and F. Raes. 2007. "Making integration of adaptation and mitigation work: mainstreaming into sustainable development policies?" *Climate Policy* 7(4): 288–303.

Swyngedouw, E. 2004. *Social Power and the Urbanization of Water: Flows of Power*. Oxford; New York: Oxford University Press.

Swyngedouw, E. 2010. "Apocalypse forever? Post-political populism and the spectre of climate change." *Theory, Culture and Society* 27(2–3): 213–32.

Taylor, M. 2009. "Displacing Insecurity in a Divided World: global security, international development and the endless accumulation of capital." *Third World Quarterly* 30(1): 147–62.

Thomas, D., and C. Twyman. 2005. "Equity and justice in climate change adaptation amongst natural-resource-dependent societies." *Global Environmental Change* 15(1): 115–24.

Tol, R., T. Downing, O. Kuik, and J. Smith. 2004. "Distributional aspects of climate change impacts." *Global Environmental Change* 14(3): 259–72.

Turner II, B. L. 2010. "Vulnerability and resilience: Coalescing or paralleling approaches for sustainability science." *Global Environmental Change* 20(4): 570–76.

UNDP. 2007. *Human Development Report 2007/2008: Fighting Climate Change – Human Solidarity in a Divided World*. Basingstoke, UK: Palgrave Press.

United Nations Framework Convention on Climate Change. 2007. *Climate Change: Impacts, Vulnerabilities and Adaptation in Developing Countries*. Bonn: United Nations.

Verner, D. (Ed.). 2010. *Reducing Poverty, Protecting Livelihoods, and Building Assets in a Changing Climate: Social Implications of Climate Change for Latin America and the Caribbean*. Washington, DC: World Bank.

Williams, R. 1980. *Problems in Materialism and Culture*. London: Verso.

Wisner, B., P. Blaikie, T. Cannon, and I. Davis. 2004. *At risk: Natural Hazards, People's Vulnerability and Disasters*. London: Routledge.

World Bank. 2006a. *Managing Climate Risk: Integrating Adaptation into World Bank Group Operations*. Washington DC: World Bank Group.

World Bank. 2006b. *Overcoming Drought: Adaptation Strategies for Andhra Pradesh*. Washington, DC: IBRD.

World Bank. 2007. *World Development Report 2008: Agriculture for Development*. Oxford: Oxford University Press.

World Bank. 2008. *Climate Change Impacts in Drought and Flood Affected Areas: Case Studies in India*. Washington, DC: IDRC.

World Bank. 2009. *World Development Report 2009: Development and Climate Change*. Washington, DC: World Bank.

World Bank. 2010. *The Economics of Adaptation to Climate Change*. Washington, DC: IBRD.

World Bank. 2013. *Turn Down the Heat: Climate Extremes, Regional Impacts, and the Case for Resilience. A report for the World Bank by the Potsdam Institute for Climate Impact Research and Climate Analytics*. Washington, DC: World Bank.

Zacune, J. 2011. *World Bank: Catalysing Catastrophic Climate Change*. Amsterdam: Friends of the Earth.

4 Power, inequality and relational vulnerability

In a world of deep and enduring inequalities, there is a growing recognition that the impacts of climatic change will not only permeate established social fractures but are set to wrench them further apart. In so doing, as the UNDP has argued, climate change will reinforce the vast disparities between the 'haves' and the 'have nots' both within and across countries, creating what it provocatively terms an 'adaptation apartheid' (UNDP 2007: 2). In response, a more concerted attempt to grapple with the interface between social inequality and vulnerability to climatic change is slowly emerging within the adaptation literature. A World Bank-sponsored publication, for example, recently emphasised how vulnerability to climate change is socially differentiated along pre-existing lines of social, material and political inequality. For its authors, "social as well as political-economic structures will influence the impact of climate shocks and trends. People that contend with multiple inequalities will be most vulnerable to climate change" (World Bank 2010: v). This is a critical point. In exposing the deeply stratified nature of the contemporary global political economy, the intersection between climatic change and social inequality asks important questions of the adaptation paradigm. To what degree can the concept of adaptation – with its emphasis on overcoming a first-order rupture between humans and their environment – convincingly capture the ways in which societies are themselves inherently divided along lines of class, gender, ethnicity and other social fractures?

To address this question, I return to the main concepts present in the discourse of climate change adaptation, specifically those of resilience and vulnerability, to ask how issues of inequality and power can be represented through such frameworks. To do so I examine two contrasting perspectives in turn – resilience and human security – and discuss the contributions and limitations of each with specific reference to agrarian environments. Notwithstanding their noted strengths, I argue that these perspectives provide few analytical tools by which to understand vulnerability in relational terms: i.e. how the relative security of some can be produced through a series of socio-ecological relationships that reproduce the relative vulnerability of others. As a result, despite their radical differences, neither framework is well attuned to understand how power is deeply embroiled in the experience of climatic change.

Through examples drawn from agrarian environments in South Asia, I emphasise how various actors are positioned within a complex and hierarchical field of socio-ecological relations. In order to pursue their livelihoods, subordinate groups repeatedly require access to assets controlled by dominant groups. These dependent relationships create a stratified human landscape in which the risks and opportunities presented by climatic change are unequally distributed and in which the vulnerability of the marginalised serves as a buttress to the security of others. In such context, climate change manifests itself not as an external environmental threat but as a shift in socio-ecological relations through which households reproduce themselves as part of a broader social division of labour. Climatic change, therefore, does not simply impact upon social groups differently. Rather, it expresses itself through the relations of class, caste and gender that structure agrarian environments. In profiling the effects of drought upon agrarian relations in Maharashtra and Karnataka, I use this approach to question the meaning of adaptation in stratified agrarian landscapes wherein climate change opens new accumulation strategies for rural elites.

Systems thinking and the politics of resilience

Although it has not always been recognised by protagonists, one of the striking facets of the discursive terrain of climate change adaptation is the preponderance of the system concept. This concern with systemic dynamics appears to be hardwired into the rubrics of adaptation. As framed by the IPCC, for example, the 'holy trinity' concepts of vulnerability, adaptive capacity and resilience are all defined in relation to a 'system' as the unit of analysis. Vulnerability, for example, is defined as the degree to which a system is susceptible to, and unable to cope with, adverse effects of climate change; adaptive capacity is the ability of a system to adjust to climate change; and resilience is the ability of a social or ecological system to absorb disturbances while retaining the same basic structure and ways of functioning. This focus on systems analysis is well represented in academic work too. A collection of leading adaption analysts, for example, note that the academic communities concerned with both resilience and vulnerability "have an overriding concern with the response of systems to stress or perturbations" (Miller et al. 2010: 11). Although it might be contended that this is simply the result of incorporating the terminology of climate and sustainability science – in which system metaphors abound – it is important to note that such conceptual lexicon brings its own political baggage.

The actualisation of systems analysis within climate change adaptation is most clearly tied to the idea of resilience. For such traditions, the system concept serves as an analytical device to represent a relatively durable set of relations between distinct elements that form part of an integrated whole. These elements, moreover, are seen to enjoy a functional coherence that provides the basis for their durability. An ecosystem, for example, can be defined as "a functional system of complementary relations between living organisms and their environment, delimited by arbitrarily chosen boundaries, which in space and time appear to maintain a steady yet dynamic equilibrium" (Gliessman 2007: 23). For resilience perspectives, this system metaphor can be adequately stretched to include the social dimensions of ecosystem usage. For such purposes, resilience theorists have gravitated towards the idea of 'socio-ecological systems' in which ecosystems and social systems are seen as coupled parts of a greater whole. This latter concept seeks to provide a basis on which to model how ecosystems and the social systems that use and depend on them are inextricably linked and correspond with each other through a series of feedback linkages (Folke et al. 2010: 2).

Building from such notions of functional coherence and dynamic equilibrium, the question of resilience seeks to ask what happens when such systems are faced with externally generated shocks that have the potential to disrupt systemic dynamics. As proposed by American ecologist C. S. Holling (1973), the idea of resilience initially emerged as a concept to conceptualise the degree to which ecosystems could absorb outside stress while retaining their essential form and functions. The idea of resilience, therefore, intends to describe "the relationship between the system under observation and externally induced disruption, stress, disturbance or crisis" (Lang 2010: 16). On this measure, resilient systems were characterised as ones with high adaptive capacity that could "absorb disturbance and reorganise while undergoing change so as to still retain essentially the same function, structure, identity, and feedbacks" (Adger 2006). At the same time, the resilience concept also was applied as a measure of how quickly and effectively a system can regenerate, and return to its core functions, following a disturbance (Folke 2006; Miller et al. 2010). In this respect, resilience is seen as a kind of systemic property that is derived from the quality of the interactions between different components of the system. Resilient systems can absorb external shocks either without losing their fundamental structure and functions or by quickly resuming them following a period of disturbance. With contemporary climatic change perceived precisely as a form of external stresses and shocks, it is evident why the resilience concept has had such wide resonance within both academic and policy circles.

There are, of course, contending perspectives on how resilience should be conceptualised and to what units of analysis the underlying systems metaphor can apply. Within the policy literature, the gravitational pull has been towards a relatively simplistic model that emphasises resilience as a return to a previous state following an external shock. This so-called engineering resilience fits in well with institutional logics aimed at technocratic logic of preservation. For policymakers, the value of the resilience concept is its fungibility: the underlying idea of a system is eminently substitutable with a surrogate term of convenience, including households, countries, cities, a region, an industry, a community or a population. Within debates on sustainable development, for example, the system metaphor was often translated into an ideal-type notion of agrarian or natural resource-based communities in the global South (Leach, Mearns and Scoones 1999). Here the idea of 'community' served as a simple replacement for the notion of 'system', often combined with the idea of 'traditional knowledge', to provide a timeless, functionalistic glue that ensured the coherence of ecosystem and human interactions. Such constructs drew sharp critique from numerous authors within political ecology who rebuked the romanticised notions of community that underlay these analyses and the artificial nature of the boundaries it presupposed (Robbins 2000). Communities are, in practice, often heavily stratified by social divisions – from class to gender to caste – that are differentially integrated into forms of production and modes of political control that stretch far beyond the boundaries of the locality (Watts 2004).

In prising open the hidden abode of the community, the suitability of the system metaphor for the analysis of complex socio-ecological dynamics is strongly questioned. The idea of modelling social dynamics according to concepts derived from ecosystems analysis implicitly tends towards structural functionalism, in which human activities are conceptualised as intrinsically drawn towards homeostasis conditions that maintain systemic parameters. The idea of adaptive cycles, for example, appears a mechanical imposition of equilibrating systemic forces upon human actions that does not appear to provide a convincing account of actual social dynamics (Pike, Dawley and Tomaney 2010; MacKinnon and Driscoll Derickson 2013).

In the face of these kinds of critiques, resilience thinkers sometimes maintain that the concept is value-neutral in the sense that it simply observes and describes systems over time to see what is stable or persistent and what is not (Adger cited in Leach 2008: 5). Such authors, moreover, rightly emphasise the more sophisticated modelling of socio-ecological systems proposed originally by Holling. The latter sought to break decisively with the simplified steady-state notion present in single-equilibrium models and consider resilience as a property of systems with multiple points of functional attraction. As subsequently developed by authors such as Carl Folke (Folke and et al. 2002; Folke 2006), this perspective emphasises that systems tend to move between multiple equilibrium states and, therefore, display dynamic qualities in which a system is constantly changing through a process of perpetual reorganisation. Should a system be pushed through external shock outside the thresholds of its existing multiple-equilbria, it would shift to a new domain of attraction entirely with qualitatively different systemic functions and structures. This, for example, could entail a transition from semi-arid grasslands to a desert, the death of a coral reef or the salination of agricultural land. Resilience, therefore, is a measure of how a dynamic system – with coupled social and ecological dynamics – can remain within the same basin of attraction despite potentially severe external stresses.

Notwithstanding the importance of this dynamic idea of resilience, it suffers from some of the same problems when it comes to understand the political dimensions of socio-ecological dynamics. Like neoclassical economics, whose practitioners also project the illusion of holding value-free positions, a number of normative assumptions already are engrained into the concept of resilience at a foundational level. For resilience theorists, these normative assumptions are precisely engrained in the tendency to presume that systems are the appropriate heuristic device through which to model social and ecological dynamics. Through the lens of resilience thinking, the world is composed of multiple, interacting systems, each with their own functional coherence and equilibrium points. This presupposition is then used to order the world by drawing boundaries around such presumed systems and ordering them according to internal dynamics and presumed external forces. As a consequence, much resilience thinking, including that basing itself on Holling's work, assumes precisely what it would first need to demonstrate: namely, the existence of systems with relatively defined parameters, functional coherence and persistence over time.

Unfortunately, resilience authors tend not to establish the appropriateness of systemic modelling but assume a priori the presence of socio-ecological systems that possess such characteristics. They then smuggle a series of value judgements silently into their analysis about what counts as systemic dynamics and where the boundaries of a system are. To create the 'socio-ecological system' to which resilience can be attributed, for example, requires the analyst to designate a series of boundaries to separate out the elements to be included as part of the dynamic system and those that are viewed as external stimuli or shocks. Decisions over what forces are 'internal' and 'external' to systemic dynamics are inherently value-laden, yet this is rarely acknowledged within the paradigm. In designating the boundaries necessary to build their systems, resilience perspectives intrinsically tend to relegate socio-political processes that cannot be easily modelled according to notions of functional coherence to the status of 'externalities' or 'context'. This has the effect of relegating questions of power, conflict, contested interests and social struggles to an exterior realm where they can be held apart from the purity of systemic dynamics. To do otherwise would be to recognise the problems of using a system concept for explaining complex socio-ecological dynamics.

Consider, for example, the 15 case studies of social-ecological systems across continents and societies provided by Walker et al. (2006). In each, resilience is discussed as something that appears intrinsic to a localised system of ecosystem dynamics and human resource management in a way that is curiously dissociated from the wider dynamics of global socio-ecological metabolisms. The latter are presumed to belong to other systemic dynamics operating at different scales and, therefore, are acknowledged but not formally integrated into analysis. The end result is that such resilience perspectives are conspicuously shy to explore the complex spatial and temporal dimensions inherent to the production (and destruction) of nature in contemporary capitalism. While the attention to ecological timescales in the paradigm is well taken, its focus on social temporalities and processes appears incredibly thin. If the idea of a socio-ecological system was to be truly holistic, then ecological dynamics must be understood in the context of their synthesis with the temporal dimensions of social processes and, specifically, capitalist time-space in which capitalism has an inherent tendency to shrink space and compress time as a means to increase the turnover time of capital (Harvey 1999). Yet within resilience analysis, the wider dynamics of capitalist commodity exchange are portrayed shallowly as forces external to the arbitrarily constructed bounds of socio-ecological systems rather than integral elements of these very relations. At times it seems that resilience analysis was curiously written for a pre-capitalist world.

Analysts worried about such concerns would find little comfort in the kinds of case studies undertaken within much of the ecological resilience canon. The general framework for understanding system thresholds across scales and domains presented by Kinzig et al. (2006), for example, represents systemic change as driven by ecological, economic and socio-cultural dynamics. This, of course, seems an important step forward. Yet such analytical categories are composed without the presence of any conflicting interests or questions of power. The outcome is to produce a disturbingly thin account of socio-ecological dynamics that lends itself to technocratic governance by expunging the actual messiness and conflicts of social life from the model. As Cote and Nightingale (2012) note, resilience authors betray the weighty influence of a 'modelling culture' that is preoccupied with determining ecological outcomes at the expense of conceptualising power and culture that cannot be reduced to such frameworks.

This leads to an approach that frequently trips over its own limitations. The account of resilience in northern Afghan pastoral management systems by Jamila Haider et al. (2012), for example, is beset with such problems. At the outset, the communities in question are presumed to be homogenous bodies that collectively manage pasture in ways that react to external forces. That there might be relations of power, domination, resistance and acquiescence within such communities, and that those shifting dynamics might be pivotal for understanding the relative security and insecurity of varied groups and their relationship to the wider lived environment, is held outside the dynamics of the assumed system. There is no reflection, for example, on the gendered divisions of labour operating within these communities or of internal differences in control over land tenure, political power or other essential determinants of well-being. The community simply appears as a hidden abode. At the same time, there is no consideration of how such 'socio-ecological systems' are integrated into broader flows of materials, energy, commodities and finance. Despite the centrality of market dynamics to pasture management (see Chapter Eight), the latter are arbitrarily placed outside the designated socio-ecological system as an external force because it is impossible to model them in terms of equilibriums and functional coherence. To maintain the idea of a systemic dynamic, it is, therefore, necessary to externalise a series of socio-ecological relations that are fundamental to the dynamics of the lived environment in this region of Afghanistan.

While resilience theorists sometimes admit that system boundaries are arbitrary, they appear far less aware that they are inherently political. How analysis differentiates between what is considered internal and external has manifold normative ramifications for the core questions that bedevil resilience analysis: namely, what is being sustained, for whom and on what temporal horizons? It is precisely this arbitrariness of systemic boundaries and the unwillingness to unlock questions of power that makes the resilience perspective so amenable for political purposes aimed at safeguarding the status quo. Ultimately, what analysts may term 'resilient systems' can often be highly inequitable socio-ecological orders precisely because the qualities identified as 'resilience' are a function of engrained power relationships that institutionalise the displacement of risk and power between social groups. The pursuit of resilience, therefore, cannot be separated from the pursuit of stabilised systems of power. Far from a 'systemic property', resilience might instead be considered as an institutionalisation of power differentials in which an uneven distribution of risks and rewards are built into the socio-ecological relations of a lived environment. In such cases, there would be a need for 'anti-resilience' thinking: resilience might maintain a set of socio-ecological relations relatively stable but at significant costs for specific social groups. In short, one person's resilience may well be another's subjugation, and what is termed resilience might be part of the problem, not its solution. Resilience theory itself, however, appears to give little purchase to exploring those dynamics.

Human security approaches

Despite the problems that resilience perspectives have in both theorising power and recognising their own normative assumptions, other accounts engage closely with these issues. Drawing upon older debates within the field of hazard and disaster studies, a critical wing of the adaptation literature has sought to explicitly reinsert questions of power and inequality into the issue of environmental change in general and climatic change in particular (Kelly and Adger 2000; O'Brien 2006; Barnett and Adger 2007; Matthew et al. 2010; O'Brien, St. Clair and Kristoffersen 2010; Pelling 2011). This literature has argued that vulnerability is not a systemic property but a socially contingent phenomenon that is a product of an existing terrain of social inequality and marginalisation. Such pre-existing inequalities stratify both the impacts of climate change and possibilities to respond. In response, finding a way not only to integrate but also to foreground the social dimensions of climate change appears as a necessary step for public policy formation that could pay more than lip service to questions of equity and sustainability (Kelly and Adger 2000). In particular, by failing to engage questions of social differentiation within target populations, the implementation of technocratic approaches based on engineering resilience risk creating situations in which local elites capture the benefits of interventions and the marginalised end up excluded from or negatively impacted by such processes (Adger 1999; Adger, Arnell and Tompkins 2005; O'Brien and Leichenko 2007). In this sense, the cure of adaptation potentially can be worse than the curse of climate change exposure, a point stressed by Marino and Ribot (2012).

An important subset of this critically orientated literature has consolidated around the concept of human security (Matthew et al. 2010; O'Brien, St. Clair and Kristoffersen 2010). The latter came to prominence in the 1990s under the auspices of the UNDP as the basis on which to build a 'people centred' development paradigm that could challenge neoliberal fixations with economic growth and state-centric approaches to international security that dominated international policymaking in the post-Cold War era. Posed as an umbrella framework that could capture a diverse spectrum of human well-being, human security became a conceptual centrepiece for numerous scholars and practitioners seeking to reinforce the importance of public policy for securing individual safety from chronic threats such as hunger, disease, and repression alongside protection from sudden and hurtful disruptions in the patterns of daily lives (UNDP 1994: 23). Strongly influenced by the work of Amartya Sen, public policy was envisaged as a platform to enable individuals and households to develop their capabilities so as to more fully participate in and benefit from the processes and events that shape their lives (Sen 1981, 1985; UNDP 1994).

Within the context of climate change adaptation, two immediate advantages of the human security framework can be discerned. On the one hand, the concept of human security breaks from the notion of a bounded system as the unit of analysis and, therefore, helps to avoid the structural-functionalism latent within the resilience paradigm. On the other, it allows analysts to flag precisely the equity concerns that tend to get smothered in technocratic analysis. As such, the human security concept is seen to provide a clear guide for the conceptualisation of climate change adaptation wherein it offers a distinct normative goal for policy (Vogel and O'Brien 2004). Adaptation, for the human security framework, is the movement towards a condition wherein individuals and communities have the options necessary to end, mitigate or adapt to risks to their human, environmental and social rights. At the same time, the framework accentuates the importance of participation and choice within such a process of adaptation. A fundamental prerequisite of human security is that individuals and communities have the capacity and freedom to exercise these options and actively participate in attaining them (GECHS 1999).

That this formulation clearly reflects the influence of Amartya Sen is not surprising given that Sen's analysis of entitlement failure made a seminal contribution to food security and disaster studies that helped frame the original human security paradigm (Sen 1981; UNDP 1994). In his theory of entitlement failure, Sen demonstrated how famines were not caused by a decline in the supply of food per se. Rather they were formed through a change in social conditions that prevented numerous individuals from turning their 'endowments' (assets and resources) into 'entitlements' (a bundle of commodities and services) that could prevent starvation (Sen 1985). This analysis proved highly influential for the human security approach because it provided a coherent analytical framework in which to conceptualise the relationship between individuals and their social context. While Sen focused on market exchange as the primary means of mapping endowments to entitlements, the latter also could be achieved through claims upon the state for public goods or from collective property resources, such as community-managed forests to common grazing lands (Devereux 2001).

An emphasis on such heterogeneous forms of entitlement mapping allows human security-orientated scholars to strengthen their understanding of the social dimensions of vulnerability in a manner that partially dovetailed with important work within early political ecology (Watts and Bohle 1993; Blaikie et al. 1994; Bohle, Downing and Watts 1994). Vulnerability, as noted in the previous chapters, is commonly seen as the product of an interaction between an external side of risks, shocks and stress to which an individual or household is subjected and an internal side which is defencelessness, meaning a lack of means to cope. The human security paradigm emphasises precisely this latter dimension, stressing how existing social differentiation strongly configures the degree of exposure and the relative adaptive capacity of different social groups. As an early contribution noted:

> [A]ny analysis of vulnerability must consider the 'architecture of entitlements', the social, economic and institutional factors that influence levels of vulnerability within a community or nation and promote or constrain options for adaptation.
>
> (Kelly and Adger 2000)

In posing these questions, human security-orientated approaches help provide conceptual clarity regarding the relationship between poverty and vulnerability wherein the two can be understood as distinct but intimately related concerns. Poverty is described as a condition in which mutually reinforcing social, economic and political processes undermine the capability of households to achieve valued 'functionings', such as secure employment and self-respect (Sen 1999). In contrast, vulnerability highlights a situation in which an individual, household or community lacks the necessary adaptive capacity to adjust to shocks or hazards in a way that maintains livelihoods and other social functions (Adger 2006; Barnett and Adger 2007). This analytical separation avoids collapsing the two conditions into each other: the poorest members of society are not necessarily the most vulnerable and vice-versa, although the two conditions are in practice frequently mutually reinforcing (see also, Chambers 1989).

One of the advantages of this framework is its explicit policy relevance. Eriksen and O'Brien (2007), for example, suggest that development and adaptation interventions should focus precisely on the overlap between poverty reduction and vulnerability reduction in an attempt to forge what they term 'sustainable adaptation' (Eriksen et al. 2011; see also, Brown 2011). The latter would be a form of governmental action that seeks to strengthen the adaptive capacity of the poor while addressing the structural causes of vulnerability. Taking these two facets together, and combining them with Sen's emphasis on agency as a fundamental aspect of development, the human security paradigm offers a laudable normative goal in the context of climatic change: human security is attained when people and communities have the capacity to manage stresses to their needs, rights, and values, and therein minimise their vulnerability to adverse climate change effects while simultaneously expanding their capabilities to improve welfare. These sentiments are, of course, well taken and unobjectionable. By emphasising the role of social differentiation in unevenly distributing the impacts of climate change while simultaneously differentiating the relative abilities of different social groups to respond, the paradigm suitably critiques the notion of technical fixes to climate change adaptation. As such, it rightly emphasises how various social inequalities that crystallise over divisions such as gender, ethnicity, age and class, serve to structure vulnerability.

The problem of power

While rightly foregrounding these key questions, it is notable that such authors remain remarkably ambiguous over how to engage them. Power is frequently invoked as an underlying determinant of inequality and vulnerability, yet its conceptualisation remains superficial. Neil Adger, for example, notes that vulnerability "is driven by inadvertent or deliberate human action that reinforces self-interest and the distribution of power in addition to interacting with physical and ecological systems" (2006: 270). While this statement usefully highlights the ways in which opposed interests and power differentials shape the experience of climatic change, quite how these relationships produce and reproduce vulnerability is left unsystematised (see also, Adger et al. 2009; Adger, Lorenzoni and O'Brien 2010). This ambiguity about how to ground vulnerability within a broader social theory is reflected in a wide array of human security-influenced literature on environmental change in general and climatic change in particular (Thomas and Twyman 2005; O'Brien 2006; Barnett and Adger 2007; Eriksen and O'Brien 2007; O'Brien and Leichenko 2007; Barnett, Matthew and O'Brien 2010). Although such progressive adaptation authors are clearly aware that power keenly matters, they often seem reluctant to tackle such thorny questions head-on by delving into the kinds of critical theory that seek to systematically engage such issues.

My concern here is that, in sidelining these expressly important dynamics, the human security approach lends itself to a peculiar patchwork framework wherein the paradigm tends to identify the importance of power and inequality without providing clear analytical tools by which to analyse them. This, in turn, leads to a selective eclecticism in the analysis of adaptation that offers glimpses of a more coherent framework without fully delivering it. For example, the human security analysis of the impact of Hurricane Katrina in New Orleans provided by Victoria Basolo (2010) traces out how social differentials on the basis of race, age and income stratified vulnerability to flooding without interrogating why and how such inequalities have been produced or reproduced in the first place. Similarly, an interrogation of vulnerability to extreme weather among urban slum dwellers by Laura Little and Chris Cocklin (2010) maps out the vectors of vulnerability without considering the political economy of why urbanisation in this specific form is occurring and the power relations that reproduce such uneven control over urban space. An insightful cartography of urban inequality, therefore, is provided without a correlate analysis of its causative dimensions within the production of the built environment.

Viewed from a political ecology perspective, there emerges an unnerving sense that this branch of the adaptation paradigm has caught the tiger by the tail and remains unsure what to do with it. Such authors recognise that what they term climate change adaptation is inherently a question of power because 'adaptation' hinges upon the ability to transform prevailing socio-ecological relations. This entails that 'adaptation' – or how specific social groups can alter their position within a lived environment to provide relative security – is only forged through relationships with other social groups. Household livelihoods and their physical security are shaped by a particular constellation of socio-ecological relations in which, as Erik Swyngedouw puts it, "social actors strive to defend and create their own environments in a context of class, ethnic, racial and/or gender conflicts and power struggles" (2006: 106). Yet an analysis of power is frequently held at arms length in the human security paradigm, wherein it tends to be reduced to a static notion of inequality that requires appropriate social policies. The overall result is a tendency to create ever-more detailed descriptions of vulnerability while remaining studiously abstinent on the underlying causes (Hinkel 2011). Worryingly, this can reinforce an approach to policymaking that seeks to address vulnerability simply in terms of its symptoms. 'Adaptive capacity', for example, is often discussed normatively in terms of vulnerable groups that need empowering without specifying how their disempowerment has been created historically and is reproduced on a quotidian basis (World Bank 2010; Eriksen et al. 2011).

For some, such as Des Gasper, this unwillingness to engage systematically with the socio-ecological determinants of vulnerability is one of the enduring qualities of the human security paradigm. In his reading, the concept of human security is simply a useful entry point for investigation. If forms a heuristic tool for consolidating a discourse and action around a shared series of concerns (Gasper 2010). This paean to eclecticism, however, is to make a virtue out of the approach's deficiencies. While human security analysis helps us understand how vulnerability is expressed, it remains gravely weakened by its reluctance to engage the power relations that produce and reproduce vulnerability over time. As a result, the paradigm occupies an anxious position as the critical wing of a profoundly managerialist discourse. It strives to cast a wide analytical net so as to gather the essential coordinates of vulnerability while simultaneously avoiding catching the socio-ecological power relations that create them. Such tepidness is, in part, because a focus on power does not lend itself easily to the kinds of managerial solutions that animate the policy sphere that human security perspectives seek to engage.

In this manner, the concerns of human security analysis often align closely with the neoinstitutionalist perspective of the World Bank. As noted in the introduction above, the Bank has been increasingly cognisant of the social dimensions of climate change. In a publication it recently sponsored, the authors emphatically pronounce that: "A socially just approach to climate change emphasizes the need to empower those most vulnerable by amplifying their voices, improving institutional capacity to better service their needs and reducing existing inequalities in access to assets and resources" (World Bank 2010: 5). Such a narrative is evidently an important advancement on the earlier technocratic approaches supported by the bank (World Bank 2008, 2009). The implications of this mainstreaming, however, should give human security theorists pause for thought because it closely mirrors what occurred within debates on global poverty a decade earlier. Whereas human security scholars are reluctant to situate their paradigm within a broader theory of social relations, the World Bank suffers from no such hesitation.

On the contrary, concerns around vulnerability can be folded seamlessly into its revised neoliberal framework of the 'post-Washington consensus', with a strong fixation on addressing vulnerability through expanding market access and promoting good governance (Fine 2001; Harrison 2004).

The World Bank's *World Development Report 2000/1*, for example, set out a framework for conceptualising poverty and vulnerability that borrowed key concepts from the human security lexicon. Poverty, it stipulated, is best conceptualised as the interaction between a lack of income and assets, a sense of voicelessness and powerlessness in the institutional realm and the vulnerability to adverse shocks (World Bank 2001). Posing the question in these terms, however, did not force a paradigm shift in the Bank's policy agenda. Rather, the discursive parameters of human security analysis were incorporated within a neoliberal framework of livelihoods analysis that builds out from the household in much the same way that mainstream economics unfolds its models from the unit of the rational individual. In this perspective, households are conceptualised as holders of specific assets – represented in terms of physical, human and social capital – that they utilise in the context of potential employment or entrepreneurial opportunities provided by markets. As Naila Kabeer has argued, such approaches are anchored in "standard neoclassical assumptions about individuals and households seeking to maximise their welfare in environments characterised by imperfect information and diverse forms of risk" (Kabeer, Mumtaz and Sayeed 2010). Indeed, the framework reaches its logical yet entirely fetishised endpoint by conceptualising the social context that conditions the agency of households as simply one more 'asset' called social capital (van Dijk 2011).

From this perspective, household poverty and vulnerability to external shocks are produced by constraints upon market integration and barriers to accessing public goods, including education and healthcare. These stem from social and institutional factors outside the realm of the market, such as gender, ethnic or caste discrimination, corruption, lack of information and political voicelessness that obstruct poor people from effectively exchanging assets in the market or accessing the public goods necessary to improve their human capital (World Bank 2000, 2001, 2004, 2010). Poverty and vulnerability, therefore, are conceptualised as the outcome of institutionalised social exclusion that deprives particular social groups of the ability to participate fully in economic and political life. These inequalities of opportunity tend to compound one another, leading to an inability of some households to build endowments and to map them into entitlements. The end result is entrenched vulnerability and chronic poverty (World Bank 2004, 2005).

Several policy conclusions unfold from this conceptual base. Most pertinently, in the words of the World Bank: "Societies have to help poor people overcome the obstacles that prevent them from freely and fairly participating in markets" (2000: 61). This can be achieved primarily by strengthening institutional forms through 'good governance' measures that promote accountability, transparency and pluralism. As the bank itself notes, in conditions of socially differentiated vulnerability, poor institutional frameworks "can severely limit the ability of individuals and households to build adaptive capacity" (World Bank 2010: 34). Better governance structures, it is envisaged, can create more robust market structures and help ensure both livelihood diversification and the accumulation of 'social capital'. Simultaneously, it can also ensure a more pluralistic planning of technical interventions to promote the resilience of physical infrastructure while expanding the scope and targeting of social safety net measures. The upshot of such collective transformation is to promote a greater degree of adaptive capacity, defined as "the ability of institutions or people to modify or change characteristics or behaviour so as to cope better with existing or anticipated external stresses from climate" (World Bank 2010: 30).

There are, of course, multiple ways in which institutionalised discrimination, poor governance and corruption greatly compromise the ability of particular households or social groups to achieve the kind of conditions highlighted by the notion of human security. As Leary et al. note in their review of adaptation across contexts, the "persons who are most vulnerable to climate risks are often socially and politically marginalized and therefore unable to influence government to act in their interest" (2010: 13). To understand such disempowerment, however, we need to ask about the broader structures of domination that order societies from the bottom up. Focusing narrowly on governance issues leads to impoverished conceptualisations of poverty and vulnerability that offer no basis from which to understand how markets themselves can serve as conduits through which entrenched vulnerability is produced and reproduced. It is not just that such approaches fail to question the stratified distribution of assets between social classes and groups, although this is problematic enough. Rather, they fail to investigate the essentially *relational* nature of assets in which possession of particular commodities provides a source of social power over the conduct of others. Notably, many of these ambiguities in the neoinstitutional and human security approaches can be traced back to the work of Sen, whose perspective made only a partial break from the individualist microfoundations of mainstream economics (Fine 1997; Selwyn 2011).

Markets, institutions and vulnerability in agrarian environments

Consider, for example, O'Brien and Leichenko's (2007) human-security orientated account of agrarian vulnerability under conditions of climate change and economic liberalisation in rural southern India and South Africa. Their approach carefully notes the ways in which these policy shifts and environmental changes create very differing risks and opportunities between social groups based on existing inequalities of assets and opportunities. To this end, the authors advise that a range of governmental and non-governmental institutional support, ranging from crop-storage infrastructure to marketing aids alongside more robust public provisioning, are vital to ensure that both large and small farmers are able to adapt to climatic change. While there is little doubt that the forms of exclusion highlighted by O'Brien and Leichenko are important determinants of human insecurity, the authors provide no analysis of how climate change impacts are filtered through key social structures, such as property relations, that structure the access to and use of land, labour, water, credit and markets. In short, they present an account of changing socio-ecological relations under climate change that excludes the key sources of security and vulnerability within these agrarian environments. By conducting an analysis focused largely on inequalities of market access and public goods, the approach risks excluding important causative determinants of human insecurity, resulting in a partial set of policy recommendations.

This disjuncture occurs because poverty and vulnerability cannot be understood solely as the outcomes of inequalities in market access or public goods without first setting those entitlements within the wider socio-ecological relations that animate the agrarian environment. To do so we must consider not only patterns of social exclusion but also the mechanisms of 'adverse inclusion' through which subordinate groups are incorporated within profoundly unequal power structures (Francis 2006; Hickey and du Toit 2007; Mosse 2010). In agrarian environments, this adverse incorporation is commonly predicated upon sharply varying control over key productive assets, such as land, water and capital, between classes, genders and along ethnic or caste lines. These differentials shape not only the spectrum of potential livelihoods a household enjoys or how it may cope in conditions of stress but, more fundamentally, the social relationships they must enter to achieve these goals. In order to engage agricultural production, rural households habitually depend on a range of assets held by others, such as land, water, seeds, fertiliser, pesticides, mechanical equipment and technical knowledge. Moreover, they are also dependent on others for credit for productive activities or consumption, sources of labouring opportunities, avenues of political influence and cultural inclusion. As such, households are rendered vulnerable not simply because of a lack of assets, but because they are fundamentally dependent on other social actors to turn their existing assets into tangible livelihoods and meaningful lives.

Within such context, the types of assets a household holds shape not only the spectrum of potential livelihoods options it might pursue or how it may cope in conditions of stress but also the relationships they must enter to achieve these goals. In conditions of strong social stratification, these divisions serve to reinforce relations of power and dependency between social groups that are buttressed by material and symbolic exchanges and are embedded within cultural norms surrounding divisions of class, caste and gender (Haynes and Prakash 1991). As other actors control the productive assets necessary for the pursuit of livelihoods by marginal households, the latter habitually find themselves locked into enduring social relations that are markedly unequal and which facilitate the transfer of surpluses, either through debt, unequal exchanges or the rendering of underpaid services (Mustafa 2002; Harriss-White 2005; Mosse 2007). The study of agrarian environments illustrates this point well because, in a social division of labour where households do not possess the full means to subsistence and where key social resources such as land, labour and water have been largely commoditised, households are necessarily dependent on the commodities possessed by others in order to convert their assets into entitlements. In spite of the formal equality of market exchanges, possession of certain commodities endows their possessors with distinct social powers that allow them to profit from the less socially advantageous endowments of others. Assets are, therefore, not intrinsically valuable or powerful. They become so within a historically formed socio-ecological context that is dynamic and changing.

For the rural poor, household income flows are commonly patched together through a precarious mix of subsistence or commercial agriculture, waged labour, petty commerce and circular or permanent migrations. Such processes frequently involve engaging in a complex tapestry of social relations with landlords, traders, moneylenders, corporations, local political bodies and governmental agents. Within these relationships, marginal households typically negotiate from a position of vastly unequal material and symbolic resources. A case study in north-western Nepal provided by Onta and Resurreccion, for example, shows the complex power relations that can structure relational dependency and security. In a context of strong climate variability, and stratified by divisions of caste, gender and class, Dalit women found themselves locked into a relationship of dependency with higher-caste Lama households, wherein they were compelled to increase their labour services in return for a caste-based form of patronage necessary to gain basic food security. Simultaneously, the Dalit males found themselves dependent on Lama counterparts for accessing networks for migrant labouring opportunities (Onta and Resurreccion 2011). Dependency, vulnerability, security and surplus extraction, therefore, were intimately intertwined in complex ways.

These dynamics form the core of relational vulnerability on a local level, in which the relative security of some and the vulnerability of others are directly linked and are reproduced over time. Indebtedness is often a key lever of such relational vulnerability. In the setting of semi-arid India, for example, practices of circular migration often entail the temporary movement of household members out of the home village during the dry season to take on labouring work in regional industries such as construction, mining, quarrying, brickmaking and agricultural labour on irrigated lands. For marginal households that are often heavily indebted and lack social networks to facilitate job seeking in urban centres, it is common to take on contracted labour through an intermediary who advances wages in return for a debt to be paid off by the worker's labour over the forthcoming months (Breman 2003). The prevalence of debt traps spurred on by the irregular cycles of agrarian cash flows can feed these relationships of neo-bondage from one year to the next. They therein consolidate processes of vulnerabilisation and impoverishment over time. Concurrently, employers profit from this social construction of cheap labour, while middlemen and moneylenders reap considerable rents from ensuing cycles of over-indebtedness and repeated bondage (Picherit 2012).

Such processes, moreover, are strongly gendered. The ownership of property and the types of employment and mobility available to migrants are strongly segregated along gender lines, resulting in a markedly gendered division of labour that unequally orders the distribution of rights, responsibilities and rewards between men and women. For example, with the increasing reliance of households upon male migration for construction jobs, the agricultural labour force in many parts of southern India has become increasingly feminised (da Corta and Venkateshwarlus 1999; Garikipati 2008, 2009; Rao 2011). While women step into the employment gap created by male migration, wages paid for agricultural work are only between 40 to 70 per cent of those paid to men, and women typically occupy the worst paid segments of the agricultural labour force (Garikipati 2008). This emphasises the strongly gendered dimension of agrarian environments, in which women are expected to respond to the pressures upon rural livelihoods by increasing unpaid work on household fields alongside underpaid waged work on those of others. It is these same women, moreover, that are often expected to become the agents of rural transformation through microfinance-inspired entrepreneurial activities while policing the financial activities of their cohort to ensure sound fiscal management (Taylor 2012).

Climate change as an accumulation strategy

Once we adopt a relational approach to vulnerability, we cannot be content to talk about inequalities of market access without first analysing how households are situated within the wider relations of production and social reproduction that shape contemporary agrarian environments. To do so we must break with a familiar trope of neoinstitutional approaches that rests on a neoclassical idea of market exchange as a rational interaction between maximising individuals: namely, the idea of exclusion from markets being the primary source of vulnerability for rural households. Although such a framework accords well with the dominant idea of development as driven through market incorporation (see Taylor 2010), it tends to ignore that the majority of the rural poor are already systematically included in markets, yet these relations are ones over which they exercise very little power. We therefore need to acknowledge that markets play complex roles in agrarian environments that cannot be simply reduced to a technical device for coordinating rational exchanges or providing opportunities for livelihoods. As multiple contributions to agrarian political economy have emphasised, markets structure a diverse set of power relations that not only circulate goods between actors but simultaneously serve as a means of transferring surpluses and displacing risks between hierarchically positioned agents (Crow 2001; Jan and Harriss-White 2012). As Timothy Mitchell noted, the concept of 'the market' is a simplified image for representing the complex relations between diverse social groups including farmers, labourers, landowners, state officials, international agribusinesses, consumers and others. As a representational device, it "reduces these interrelated but very unequal concentrations of power into nominally equivalent buyers and sellers, and represents the inequality between them as the market's equilibrium" (Mitchell 2002: 227).

Markets, therefore, must be understood historically as an embedded element of the wider power relations prevailing in the socio-ecological context in which they are embedded (Taylor 2010). They reflect institutionally sanctioned forms of ownership and exclusion that are pivotal elements of relational vulnerability within agrarian environments. Consider, for example, the severe drought that afflicted semi-arid areas of Andhra Pradesh, Karnataka and Maharashtra across the winter and into the spring of 2012–13. In conditions of severely deficient rainfall, the increasingly desperate extraction of groundwater led to a depletion of the water table that undermined the wells that provide both irrigation and drinking water for villages across the region. Across both Anantapur district in southwest Andhra Pradesh and Dharwad district in neighbouring Karnataka, the scarcity of water undermined the production of fodder for cattle. This left smallholders that had invested in such assets unable to grow or purchase enough fodder to feed them. The deficit of fodder, moreover, had been exacerbated by a shift of cultivation into the mono-cropping of groundnuts, a cash crop that produces very little plant material that can serve as animal feed. Without access to fodder, smallholders witnessed a key element of their asset base starving to death before their eyes.

As a result, the drought recalibrated the field of socio-ecological relations in the region, therein affecting the power relations between differently positioned actors. Unable to provide fodder, smallholders began to sell cattle en masse and such distress sales opened up new accumulation strategies for the local merchant and landlord classes. The latter bought up large amounts of cattle from individual smallholders and transported them for sale in the regional cattle markets and then further to the cities of Chennai, and Hyderabad. Through their control of capital, credit and transportation, merchants were reportedly making profits of IR 5,000 per animal precisely while smallholders lost a significant part of their asset base (*Deccan Chronicle* 2012). Such an outcome, moreover, helped to re-establish the long-term vulnerability of smallholders within the agrarian environment and their dependence on other social classes. For those rendered vulnerable under these relations, the result is frequently a pressing need to take on further debts advanced by moneylenders, landowners or labour contractors that reinforce further the extraction of surplus between classes. Drought, therefore, acts as a periodic recalibration of agrarian social relations in which debt relations function as a fulcrum of surplus extraction.

This example may seem perhaps to be an indication of agrarian atavism – the persistence of what might be termed semi-feudal relations in a region not yet transitioned to capitalist modernity. There is nothing atavistic, however, about these accumulation strategies operating across an extensively commodified agrarian environment wherein the power to extend credit or to assume debt serves as a key determinant of access to land, water, paid labour and political power. These are precisely the type of socio-ecological relations through which climatic change will express itself across much of South Asia (see Chapters Six, Seven and Eight). For example, simultaneous to the shakedown of smallholders by livestock merchants, a market for water rapidly congealed in drought-stricken semi-arid Maharashtra that was structured by heavily differentiated access to groundwater sources. As discussed in Chapter Seven, drilling deep borewells is a costly business that requires significant capital. The resulting uneven access to water stratified an agrarian environment wilting under drought. For those who controlled the deepest borewells, water could be extracted from deep within aquifers while simultaneously depleting groundwater sources for their neighbours without such profound wells.

Having located, fixed and subsequently monopolised water, such well-adapted entrepreneurs could then service local areas at extortionate prices through a network of roaming tankers. The *Deccan Chronicle* quotes Ramesh, an auto-rickshaw driver, on the dynamics of this socially constructed scarcity:

> We don't have an option but to buy water from these private tankers. My family needs about 10 pots of water every day, Rs 50 for drinking water alone. The tankers know how to do business – they come once every 3 days to our village.
>
> (*Deccan Chronicle* 2012)

The most sizeable margins, however, are not made selling to villagers whose fields are parched and whose drinking water wells have run dry. Rather it is sales to industry that are most profitable. As veteran journalist P. Sainath (2013) notes, a vibrant accumulation strategy has opened up wherein landowners and businesses make investments in deep wells, frequently in water-scarce agricultural areas. They then extract the water – which depletes local sources for other users exacerbating scarcity – and transport it via tankers to industrial clients located in urban areas. There it can fuel a quite different socio-ecology of industrial production and set of commercial interests. Viewed in terms of accumulation strategies, the impacts of climatic change combine with the commodification of water and new forms of market-entrepreneurship to forge a brutal rationality of 'adaptation'. As one political leader interviewed by Sainath put it: "If I owned ten tankers, I'd have to pray for drought this year, too" (Sainath 2013). In a world of competitive accumulation, the meaning of adaptation as a process of "facilitating societies to take advantage of the new opportunities provided by a changing environment" takes on a quite distinct hue that appears to be entirely marginalised within the established discourse.

Public policy: to adaptation and beyond

In mapping out the notion of relational vulnerability, the above sections implicitly problematised the conceptual boundaries of climate change adaptation. Analytically, thinking of vulnerability in relational terms raises a number of core issues. First, the dividing lines central between security and vulnerability, and agency and dependence, are extremely complex. They are formed not in relation to abstract climatic threats but are situated within grounded socio-ecological relations through which lived environments are produced. Vulnerability to climatic change, therefore, must be understood in a context wherein the security of some and the relative insecurity of others are directly intertwined. This involves a focus not on static inequalities but on the dynamic socio-ecological relations in which households are situated. Climatic change is an important dynamic force within such relations, yet households do not adapt to climate change per se but to the ways in which it shifts the socio-ecological parameters through which they seek to reproduce themselves. For some actors within an agrarian environment, climate change opens new accumulation strategies in which existing assets and forms of power are heightened through changing meteorological trends. For others, it sharpens the relative power that others hold over them by changing the relational parameters in which possession of specific assets becomes more or less important.

Undoubtedly, in foregrounding questions of power, this approach tends to muddy the waters when it comes to questions of public policy. It suggests that the pressing questions that we face are not ones of 'adapting to climate change' as an external threat but rather of transforming power relations that animate agrarian environments. When we consider the implications of this perspective for public policy, it indicates that there are no simple answers, only diverse political challenges. Focusing on vulnerability as a dynamic and relational condition, however, does raise two important considerations. First, although many current development and adaptation policy measures are studiously abstinent about questions of power, they nonetheless can have significant unintended impacts on such relations, with both positive and negative outcomes for marginal groups. The recent flagship rural development programme implemented by the newly elected congress government of India in 2005, the Mahatma Gandhi National Rural Employment Guarantee Act (MGNREGA), is one such example. The stated intention of this rights-based approach, which provides each rural household the right to 100 days paid labour at the prevailing minimum wage on projects implemented by the state government, was to provide a safety net to rural households facing drastic underemployment and distress migration in the countryside. Simultaneously, the types of works undertaken by MGNREGA programs were conceptualised as dovetailing with climate change adaptation strategies. A quarter of projects, for example, relate to water conservation and water harvesting; 14 per cent relate to irrigation canals and renovation of traditional water bodies and 13 per cent relate to flood protection and drought proofing (Ministry of Rural Development 2012). The possibilities of overlaps between poverty reduction and adaptation seemed uniquely tangible (Esteves 2013).

The results, however, have been uneven. Numerous critiques have been made of the programme that highlight the prevalence of delayed payments for work, the paucity of the wages that in practice often fall below the minimum wage, the fixed limit of 100 days per household, the lack of skill-building owing to the intense manual labour of the projects, the non-involvement of communities in planning and instances of significant corruption (Mehta et al. 2011; Liu and Barrett 2013). Moreover, the projects undertaken by MGNREGA around watershed development are often left unfinished or have been inadequately completed, creating a legacy of failed wells and irrigation beset by functional inadequacies (Mahapatra, Suchitra and Moyna 2011). Notwithstanding these important deficiencies, however, the most important impacts of the MGNREGA for climate change issues may have been unintended ones. The impact of providing a guaranteed quota of working days for the poor has been to reduce their dependence on the kinds of power structures that reinforced their vulnerability. This has affected the power relations prevailing in rural areas. On the one hand it increased the bargaining power of unskilled labourers, which has pushed rural wages upwards, particularly for women workers (Carswell and De Neve 2013; Jakimow 2014). Simultaneously, it has mitigated the reliance on distress migration and can shield marginalised households from the structural necessity to enter into dependency relations, such as with labour contractors, that promote dependency over the medium term (Esteves 2013). As such, a major policy initiative that appeared to be directly aimed at 'sustainable adaptation' by addressing both poverty and vulnerability through public works to improve watersheds and other infrastructure has had its most notable effects through inadvertent consequences that impacted upon the character of power relations within the agrarian environment.

Second, this foregrounding of relational vulnerability reinforces the need to address issues of power directly within governmental programming. To turn MGNREGA into a more transformative vehicle, for example, it would be necessary to increase the labour entitlement beyond 100 days, provide work on demand to all adults, ensure timely payment of minimum wages for work done and build in the representation of communities into all aspects of the programme's implementation (Mehta et al. 2011). There would also need to be concerted efforts to ensure that established political structures of patronage and dependency are not replicated and entrenched within the operation of the programme (Corbridge and Srivastava 2013; Jakimow 2014). If successful, these means would further mitigate household dependence upon prevailing power structures, including gendered divisions of labour and caste hierarchies, wherein access to MGNREGA work provides a resource – albeit a poorly paid and often exhausting one – to women and lower castes within a brutally unequal agrarian environment (Carswell and De Neve 2013). Reinforcing access to decent employment as a fundamental and legally enshrined right within rural India, therefore, will remain an ongoing political struggle that continues to shape how climatic change is experienced as part of agrarian transformation in the decade ahead.

The kind of large-scale governmental initiatives exemplified by the MGNREGA, however, should not form the horizon of our political imagination. As noted in Chapter Seven, rural social movements in southern India have consistently articulated a notably more expansive set of proposals that take aim precisely at the underlying power relations that act as sources of security for some and vulnerability for others. Such demands include substantive land reform, protection of common property resources, shielding local agriculture from subsidised foreign imports, a restructuring of rights over irrigation, effective guaranteed minimum prices for crops, greater rights for migrant labourers and subsidised forms of credit made available to smallholders and tenant farmers. Proposals of this nature raise daunting political challenges, not least given that relations of vulnerability frequently manifest themselves in strongly stratified access to the levers of institutionalised political power. They do, however, clearly indicate what the mainstream debates isolates and externalises as climate change adaptation represents a new chapter in a broader history of conflicts and compromise over the character of socio-ecological production that are interlaced in the agrarian histories of rural South Asia (Gadgil and Guha 1993; Mosse 2003). It is to this broader understanding of agrarian transformations within their appropriate historical setting that the next chapter turns.

References

Adger, W. N. 1999. "Social vulnerability to climate change and extremes in coastal Vietnam." *World Development* 27(2): 249–69.

Adger, W. N. 2006. "Vulnerability." *Global Environmental Change* 16(3): 268–81.

Adger, W. N., N. Arnell, and E. Tompkins. 2005. "Successful adaptation to climate change across scales." *Global Environmental Change* 15(1): 77–86.

Adger, W. N., S. Dessai, M. Goulden, M. Hulme, I. Lorenzoni, D. Nelson, L. O. Naess, J. Wolf, and A. Wreford. 2009. "Are there social limits to adaptation to climate change." *Climatic Change* 93(2): 335–54.

Adger, W. N., I. Lorenzoni, and K. O'Brien. 2010. "Adaptation Now" Pp. 1–22 in *Adapting to Climate Change*, edited by W. N. Adger, I. Lorenzoni, and K. O'Brien. Cambridge: Cambridge University Press.

Barnett, J., and W. N. Adger. 2007. "Climate change, human security and violent conflict." *Political Geography* 26(4): 639–55.

Barnett, J., R. Matthew, and K. O'Brien. 2010. "Global Environmental Change and Human Security: An Introduction." Pp. 3–32 in *Global Environmental Change and Human Security*, edited by Richard Matthew, Jon Barnett, Bryan McDonald, and Karen O'Brien. Cambridge: MIT Press.

Basolo, V. 2010. "Environmental Change, Vulnerability and Disasters: The Case of Hurricane Katrina and New Orleans." Pp. 98–118 in *Global Environmental Change and Human Security*, edited by Richard Matthew, Jon Barnett, Bryan McDonald, and Karen O'Brien. Cambridge, MA: MIT Press.

Blaikie, P. M., T. Cannon, I. Davis, and B. Wisner. 1994. *At Risk: Natural Hazards, People's Vulnerability, and Disasters*. London: Routledge.

Bohle, H., T. Downing, and M. Watts. 1994. "Climate change and social vulnerability." *Global Environmental Change* 4(1): 37–48.

Breman, J. 2003. *The Labouring Poor in India: Patterns of Exploitation, Subordination, and Exclusion*. New Delhi; New York: Oxford University Press.

Brown, K. 2011. "Sustainable adaptation: An oxymoron?" *Climate and Development* 3(1): 21–31.

Carswell, G., and G. De Neve. 2013. "Women at the crossroads: Implementation of employment guarantee scheme in rural Tamil Nadu." *Economic & Political Weekly* XLVIII(52): 82–94.

Chambers, R. 1989. "Vulnerability: How the poor cope." *IDS Bulletin* 20(2): 1–8.

Corbridge, S., and M. Srivastava. 2013. "Mapping the social order by fund flows: the political geography of employment assurance schemes in India." *Economy and Society* 42(3): 455–79.

Cote, M., and A. Nightingale. 2012. "Resilience thinking meets social theory: Situating social change in socio-ecological systems (SES) research." *Progress in Human Geography* 36(4): 475–89.

Crow, B. 2001. *Markets, Class and Social Change Trading Networks and Poverty in Rural South Asia*. Basingstoke, UK: Palgrave.

da Corta, L., and D. Venkateshwarlus. 1999. "Unfree relations and the feminisation of agricultural labour in Andhra Pradesh, 1970–95." *Journal of Peasant Studies* 26(2): 71–139.

Deccan Chronicle. 2012, April 18. "Rs 5 to Quench Your Thirst." Available at: http://archive.asianage.com/bengaluru/rs-5-quench-your-thirst-796

Devereux, S. 2001. "Sen's entitlement approach: Critiques and counter-critiques." *Oxford Development Studies* 29(3): 245–65.

Eriksen, S., P. Aldunce, C. S. Bahinipati, R. D'Almeida Martins, J. I. Molefe, C. Nhemachena, K. O'Brien, F. Olorunfemi, J. Park, L. Sygna, and K. Ulsrud. 2011. "When not every response to climate change is a good one: Identifying principles for sustainable adaptation." *Climate and Development* 3(1): 7–20.

Eriksen, S., and K. O'Brien. 2007. "Vulnerability, poverty and the need for sustainable adaptation measures." *Climate Policy* 7(4): 337–52.

Esteves, T., K. Basu, I. Porsche, R. Sharma, I. K. Murthy, S. Rao, . . . N. H. Ravindranath. 2013. "Agricultural and livelihood vulnerability reduction through the MGNREGA." *Economic & Political Weekly* XLVIII(52): 94–104.

Fine, B. 1997. "Entitlement failure?" *Development and Change* 28(4): 617–47.

Fine, B. 2001. "Neither the Washington nor the Post-Washington Consensus." Pp. 1–27 in *Development Policy in the Twenty-First Century: Beyond the Post-Washington Consensus*, edited by Ben Fine, Costas Lapavitsas, and Jonathan Pincus. London: Routledge.

Folke, C. 2006. "Resilience: The emergence of a perspective for social-ecological systems analyses." *Global Environmental Change* 16(4): 253–67.

Folke, C., S. Carpenter, T. Elmqvist, L. Gunderson, C. Holling, and B. Walker. 2002. "Resilience and Sustainable Development: Building Adaptive Capacity in a World of Transformations." Edited by Report for the Swedish Environmental Advisory. Stockholm: Ministry of the Environment.

Folke, C., S. Carpenter, B. Walker, M. Scheffer, T. Chapin, and J. Rockström. 2010. "Resilience thinking: Integrating resilience, adaptability and transformability." *Ecology and Society* 15(4): 1–10.

Francis, E. 2006. *Poverty: Causes, Responses and Consequences in Rural South Africa*. CPRC Working Paper No. 60. Manchester: University of Manchester.

Gadgil, M., and R. Guha. 1993. *This Fissured Land: An Ecological History of India*. Berkeley: University of California Press.

Garikipati, S. 2008. "Agricultural wage work, seasonal migration and the widening gender gap: evidence from a semi-arid region of Andhra Pradesh." *European Journal of Development Research* 20(4): 629–48.

Garikipati, S. 2009. "Landless but not assetless: female agricultural labour on the road to better status, evidence from India." *Journal of Peasant Studies* 36(3): 517–45.

Gasper, D. 2010. "The Idea of Human Security." Pp. 23–46 in *Climate Change, Ethics and Human Security*, edited by Karen O'Brien, Asuncion Lera St. Clair, and Berit Kristoffersen. Cambridge: Cambridge University Press.

GECHS. 1999. " Global Environmental Change and Human Security (GECHS) Science Plan." *IHDP Report 11*. Available at: www.ihdp.uni-bonn.de/html/publications/reports/report11/gehssp.htm

Gliessman, S. R. 2007. *Agroecology: The Ecology of Sustainable Food Systems*. Boca Raton, FL: CRC Press.

Haider, J., A. Quinlan, and G. Peterson. 2012. "Interacting traps: Resilience assessment of a Pasture management system in northern Afghanistan." *Planning Theory and Practice* 13(2): 299–333.

Harrison, G. 2004. *The World Bank and Africa: The Construction of Governance States*. London: Routledge.

Harriss-White, B. 2005. *India's Market Society: Three Essays in Political Economy*. Gurgaon, India: Three Essays Collective.

Harvey, D. 1999. *The Limits to Capital*. London: Verso.

Haynes, D. E., and G. Prakash. 1991. *Contesting Power: Resistance and Everyday Social Relations in South Asia*. Delhi: Oxford University Press.

Hickey, S., and A. du Toit. 2007. *Adverse Incorporation, Social Exclusion and Chronic Poverty*. CPRC Working Paper No. 81. Manchester: University of Manchester.

Hinkel, J. 2011. "Indicators of Vulnerability and Adaptive Capacity: Towards a Clarification of the Science-Policy Interface." *Global Environmental Change* 21: 198–208.

Holling, C. S. 1973. "Resilience and stability of ecological systems." *Annual Review of Ecology and Systematics* 4(1): 1–23.

Jakimow, T. 2014. "'Breaking the backbone of farmers': Contestations in a rural employment guarantee scheme." *Journal of Peasant Studies* 41(2): 263–81.

Jan, M. A., and B. Harriss-White. 2012. "The three roles of agricultural markets: A review of ideas about agricultural commodity markets in India." *Economic and Political Weekly* 47(52): 39–53.

Kabeer, N., K. Mumtaz, and A. Sayeed. 2010. "Beyond risk management: Vulnerability, social protection and citizenship in Pakistan." *Journal of International Development* 22: 1–19.

Kelly, P. M., and W. N. Adger. 2000. "Theory and practice in assessing vulnerability to climate change and facilitating adaptation." *Climatic Change* (47): 325–52.

Kinzig, A., P. Ryan, M. Etienne, H. Allison, T. Elmqvist, and B. Walker. 2006. "Resilience and regime shifts: Assessing cascading effects." *Ecology and Society* 11(1): 20–43.

Lang, T. 2010. "Urban Resilience and New Institutional Theory: A Happy Couple for urban and Regional Studies." Pp. 15–24 in *German Annual of Spatial Research and Policy 2010*, edited by B. Muller. Berlin: Springer.

Leach, M. (Ed.). 2008. *Re-framing Resilience: a Symposium Report*. Brighton, UK: STEPS Centre.

Leach, M., R. Mearns, and I. Scoones. 1999. "Environmental Entitlements: Dynamics in Community-Based Natural Resource Management." *World Development* 27(2): 225–47.

Leary, N., J. Adejuwon, V. Barros, P. Batima, B. Biagini, I. Burton, . . . M. Wehbe. 2010. "A Stitch in Time: General Lessons from Specific Cases." Pp. 1–27 in Climate Change and Adaptation, edited by Neil Leary et al. London: Earthscan Press.

Little, L., and C. Cocklin. 2010. "The Vulnerability of Urban Slum Dwellers to Global Environmental Change." Pp. 77–97 in *Global Environmental Change and Human Security*, edited by Richard Matthew, Jon Barnett, Bryan McDonald, and Karen O'Brien. Cambridge, MA: MIT Press.

Liu, Y., and C. Barrett. 2013. "Heterogeneous pro-poor targeting in the national rural employment guarantee scheme." *Economic & Political Weekly* XLVII(10): 46–54.

MacKinnon, D., and K. Driscoll Derickson. 2013. "From resilience to resourcefulness: A critique of resilience policy and activism." *Progress in Human Geography* 37(2): 253–70.

Mahapatra, R., M. Suchitra, and Moyna. 2011. "A Million Opportunities Lost." *Down to Earth* (2011–12–15): 3–12.

Marino, E., and J. Ribot. 2012. "Adding insult to injury: Climate change and the inequities of climate intervention." *Global Environmental Change* 22(3): 323–28.

Matthew, R., J. Barnett, B. McDonald, and K. O'Brien (Eds.). 2010. *Global Environmental Change and Human Security*. Cambridge, MA: MIT Press.

Mehta, A. K., A. Shepherd, S. Bhide, A. Shah, and A. Kumar. 2011. *India Chronic Poverty Report: Towards Solutions and New Compacts in a Dynamic Context*. New Delhi: Indian Institute of Public Administration.

Miller, F., H. Osbahr, E. Boyd, F. Thomalla, S. Bharwani, G. Ziervogel, B. Walker, J. Birkmann, S. van der Leeuw, J. Rockstrom, J. Hinkel, T. Downing, C. Folke, and D. Nelson. 2010. "Resilience and vulnerability: Complementary of conflicting concepts." *Ecology and Society* 15(3): 11–37.

Ministry of Rural Development. 2012. "MGNREGA Sameeksha, an Anthology Of Research Studies on the Mahatma Gandhi National Rural Employment Guarantee Act, 2005, 2006–2012." Edited by Neelakshi Mann and Varad Pande edited and compiled by Mihir Shah. New Delhi: Orient BlackSwan.

Mitchell, T. 2002. *Rule of Experts: Egypt, Techno-Politics, Modernity*. Berkeley: University of California Press.

Mosse, D. 2003. *The Rule of Water: Statecraft, Ecology and Collective Action in South India*. Oxford: Oxford University Press.

Mosse, D. 2007. "Power and the durability of poverty: a critical exploration of the links between culture, marginality and chronic poverty." CPRC Working Paper, Anthropology Department, SOAS (107): 1–60.

Mosse, D. 2010. "A relational approach to durable poverty, inequality and power." *Journal of Development Studies* 46(7): 1156–78.

Mustafa, D. 2002. "To each according to his power? Participation, access, and vulnerability in irrigation and flood management in Pakistan." *Environment and Planning D: Society and Space* 20(3): 737–52.

O'Brien, K. 2006. "Are we missing the point? Global environmental change as an issue of human security." *Global Environmental Change* 16(1): 1–3.

O'Brien, K., and R. Leichenko. 2007. "Human Security, Vulnerability and Sustainable Adaptation." *Human Development Report 2007/2008*, Human Development Report Office Occasional Paper 2007/9. Geneva: United Nations.

O'Brien, K., A. L. St. Clair, and B. Kristoffersen (Eds.). 2010. *Climate Change, Ethics and Human Security*. Cambridge: Cambridge University Press.

Onta, N., and B. Resurreccion. 2011. "The role of gender and caste in climate adaptation strategies in Nepal." *Mountain Research and Development* 31(4): 351–56.

Pelling, M. 2011. *Adaptation to Climate Change*. London: Routledge Press.

Picherit, D. 2012. "Migrant labourers' struggles between village and urban migration sites: Labour Standards, rural development and politics in South India." *Global Labour Journal* 3(1): 7–29.

Pike, A., S. Dawley, and J. Tomaney. 2010. "Resilience, adaptation and adaptability." *Cambridge Journal of Regions, Economy and Society* 3(1): 59–70.

Rao, S. 2011. "Work and empowerment: Women and agriculture in South India." *Journal of Development Studies* 47(2): 294–315.

Robbins, P. 2000. "Pastoralism and Community in Rajasthan: Interrogating Categories of Arid Lands Development." Pp. 191–215 in *Agrarian Environments: Resources, Representation and Rule in India*, edited by Arun Agrawal and K. Sivaramakrishnan. Durham, NC: Duke University Press.

Sainath, P. 2013, March 27. "Tankers and the Economy of Thirst." *The Hindu*. Available at: http://www.thehindu.com/opinion/columns/sainath/tankers-and-the-economy-of-thirst/article4551597.ece

Selwyn, B. 2011. "Liberty Limited? A Sympathetic Re-Engagement with Amartya Sen's Development as Freedom." *Economic & Political Weekly* XLVI(37): 68–77.

Sen, A. K. 1981. *Poverty and Famines: An Essay on Entitlement and Deprivation*. Oxford: Oxford University Press.
Sen, A. K. 1985. *Commodities and Capabilities*. Amsterdam: North-Holland.
Sen, A. K. 1999. *Development as Freedom*. Oxford: Oxford University Press.
Swyngedouw, E. 2006. "Circulations and metabolisms: (Hybrid) natures and (Cyborg) cities." *Science as Culture* 15(2): 105–21.
Taylor, M. 2010. "Conscripts of competitiveness: Culture, institutions and capital in contemporary development." *Third World Quarterly* 31(4): 561–79.
Taylor, M. 2012. "The antinomies of 'financial inclusion': Debt, distress and the workings of Indian microfinance." *Journal of Agrarian Change* 12(4): 601–10.
Thomas, D., and C. Twyman. 2005. "Equity and justice in climate change adaptation amongst natural-resource-dependent societies." *Global Environmental Change* 15(1): 115–24.
UNDP. 1994. *Human Development Report 1994*. Oxford: Oxford University Press.
UNDP. 2007. *Human Development Report 2007/2008: Fighting climate change – Human solidarity in a divided world*. Geneva: United Nations.
van Dijk, T. 2011. "Livelihoods, capitals and livelihood trajectories: a more sociological conceptualisation." *Progress in Development Studies* 11(2): 101–17.
Vogel, C., and K. O'Brien. 2004. "Vulnerability and global environmental change: Rhetoric and reality " *Bulletin on Global Environmental Change and Human Security* (13): 1–4.
Walker, B., J. Anderies, A. Kinzig, and P. Ryan. 2006. "Exploring resilience in social-ecological systems through comparative studies and theory development: Introduction to the special issue." *Ecology and Society* 11(1): 12–17.
Watts, M. 2004. "Antinomies of community: Some thoughts on geography, resources and empire." *Transactions of the Institute of British Geographers* 29(2): 195–216.
Watts, M., and H. Bohle. 1993. "The space of vulnerability: the causal structure of hunger and famine." *Progress in Human Geography* 17(1): 43–69.
World Bank. 2000. *World Development Report 2000/2001: Attacking Poverty*. Oxford: Oxford University Press.
World Bank. 2001. *World Development Report 2002: Building Institutions for Markets*. Oxford: Oxford University Press.
World Bank. 2004. *World Development Report 2005: A Better Investment Climate For Everyone*. Oxford: Oxford University Press.
World Bank. 2005. *World Development Report 2006: Equity and Development*. Oxford: Oxford University Press.
World Bank. 2008. *Climate Change Impacts in Drought and Flood Affected Areas: Case Studies in India*. Washington, DC: IDRC.
World Bank. 2009. *World Development Report 2009: Development and Climate Change*. Washington, DC: World Bank.
World Bank. 2010. *The Economics of Adaptation to Climate Change*. Washington, DC: IBRD.

5 Climate, capital and agrarian transformations

Although climate change adaptation is a relatively new entrant into the contemporary governmental lexicon, the projection of encompassing transformations onto agrarian environments is conspicuously familiar. From the colonial period to the present, rural peoples have repeatedly found themselves as the object of state-driven projects to intensify agricultural production, more fully integrate producers into national and international divisions of labour and mediate the political discontents that such processes entail. In the present chapter, I map out how the idea and practices of climate change adaptation are situated within this wider context. I argue that there are two parallel narratives of agrarian transformation at work that, despite being held at arm's length, are duly complementary. The first is the standard rubric of adaptation introduced in the previous chapters. This narrative represents agrarian regions as spaces that are disproportionately vulnerable to climatic threats. Arun Agrawal and Nicolas Perrin, for example, highlight a catalogue of incipient risks facing natural resource-dependent households, including "droughts, famines, floods, variability in rainfall, storms, coastal inundation, ecosystem degradation, heat waves, fires, epidemics, and even conflicts" (2008: 1). The idea of adaptation is then called into being as a series of planned social adjustments to mediate such proximate climatic disturbances, including institutional changes, infrastructure building, the diffusion of new technologies and social reforms. Collectively, these adaptation processes are intended to guard against the threats posed by climatic change.

Alongside this account of adaptation as a form of risk mitigation, however, there exists a corresponding institutional narrative of agrarian transformation that coheres around climate change as a threat to global agricultural production and food security. In this framing, the object of adaptation is agriculture as an economic sector and the source of food for an expanding and increasingly urbanised global population with changing consumption habits. While the first narrative is immediate, localised and reactive, the second is explicitly posited as a vision of socio-economic transformation on a wider temporal and spatial scale. Framed in neo-Malthusian terms of population pressures and future food scarcity, it articulates the need for deep-seated changes to global agriculture that are rendered ever more pressing by the spectre of climate change:

> By 2050 the world will need to feed 3 billion more people and cope with the changing dietary demands of a richer population (richer people eat more meat, a resource-intensive way to obtain proteins). This must be done in a harsher climate with more storms, droughts, and floods. And it has to incorporate agriculture in the mitigation agenda because agriculture drives about half the deforestation every year and directly contributes 14 percent to overall emissions. And ecosystems, already weakened by pollution, population pressure, and overuse, are further threatened by climate change. Producing more and protecting better in a harsher climate while reducing greenhouse gas emissions is a tall order.
>
> (World Bank 2009: 16)

This portrayal of a pressing obligation to increase productivity in the face of an uncertain climatic future restates a familiar vision of agrarian modernisation that pre-dates current concerns with anthropogenic climate change (World Bank 1982, 2007). Indeed, for the institutions of global governance, climate change simply confirms what they already knew. Agriculture in the developing world needs to become more intensive, efficient and technologically advanced. To do so, it needs better integration into internationalised circuits of commodity exchange and, at the same time, it must lose some bodies by transferring labour out of agriculture into other sectors to create a more rational division of labour.

Although the two narratives of climate change adaptation and agrarian modernisation are not always explicitly brought together, they nonetheless share a common discursive space as part of a wider nucleus of governmental technologies that seek to represent, order and reshape the agrarian world. It is this terrain that I map out below, indicating how the adaptation–modernisation nexus – "producing more, protecting better" in the bank's terms – is situated within a conjuncture of governmental initiatives hinged to notions of food security, sustainable intensification, market expansion, environmental resilience and livelihood diversification (World Bank 2007, 2009; IFAD 2010). While the idea of adaptation focuses on the 'here and now' of climatic threats, the narrative of transformation establishes distinct normative parameters for what a suitably adapted agrarian future should look like. In this respect, by integrating the issues of climate change, rural poverty and food production, the modernisation narrative draws the practices of adaptation into a deeply political terrain that is configured by contested visions of agrarian futures.

Although the modernisation narrative presents itself in a quasi-evolutionary schematic of mutually beneficial progress, its projected transformations are mapped out over agrarian environments stratified by strikingly uneven degrees of power and privilege that are unevenly embedded within a global division of consumption that is itself profoundly unequal. In this context, the rhetoric of a mutually beneficial adaptation–modernisation nexus appears to be a problematic discursive device through which to conceptualise how unequal concentrations of control over land, water, bodies, technology and debt structure contemporary agrarian environments. In exploring such tensions, I argue that the emphasis of adaptation inverts the more pressing question that we should pose. Rather than start from the presumption of a rural topography made vulnerable by powerful climatic threats to which we must adapt, I argue that we should instead ask how we empower climate. To answer that question entails examining the politics of scale through which the localised processes that produce, constrain and sometimes obliterate rural livelihoods are intermeshed with flows of materials, energy and bodies within and beyond the agrarian realm. As I map out below, the dynamics of production, vulnerability and power in agrarian environments are therefore intimately wedded into a broader context of productive forces, corporate power and political authority. This allows us to conceptualise how the choices and compulsions rural agents experience are shaped within a field of relations that tie livelihoods to the subsidy decisions made in Washington or Brussels, the commodity futures markets in New York and Hong Kong and the corporate strategies fashioned in biotech labs in the American Midwest.

Agrarian modernisation and climate change

Over the past three decades, rural areas in much of the postcolonial world have been shaped by a series of transformations driven by the liberalisation of agriculture, the establishment of a new world trading regime for agricultural goods and the increasing promotion of market-driven forms of organising production and circulation (Goodman and Watts 1997; Bruinsma 2003; Akram-Lodhi and Kay 2009; Clapp 2012). These policy changes have helped to facilitate a new cartography of agricultural production that operates by integrating farming activities into circuits of exchange and accumulation that are progressively more networked on a global scale. Agricultural activities have, of course, long been incorporated into a global division of labour, and the experience of colonialism tenaciously drew diverse peasantries into a world market for agrarian produce (Liu 2010). The current degree of integration, however, is different in both scale and scope. As a recent United Nations report noted, world agriculture is presently characterised by tight interlinkages and economic concentration at almost all stages of production and marketing chains. This includes a consolidated role for transnational corporations that integrate agricultural trade both vertically and horizontally and, in so doing, exercise augmented power over both consumers and producers (IAASTD 2009: 7).

Although uneven across geographical space, these trends have created new ways of producing agricultural goods, affected decisions over what goods to produce and who can produce them and reshaped the networks used to source and distribute those commodities to consumers. In so doing, they have recalibrated the ways in which agrarian social relations are interweaved into wider flows and transformations of materials and energy at varied spatial scales. Transformative metabolic processes, of course, are fundamental to the basic purpose of agriculture to produce biomass of various forms for human use and consumption. Jan Douwe van der Ploeg puts this point eloquently when he notes that farming is the "ongoing interaction and mutual transformation of people and living nature" (2013: 48). As the social organisation of human labour in combination with biophysical forces to produce organic material for human use, alongside multiple other by-products, agriculture is emphatically a socio-ecological process. In agrarian environments, flows of energy and water are channelled through constructed technologies and non-human forces to drive chemical, physical and biological metabolisms that leave all parties transformed. An agrarian environment, therefore, is one of continual production, not simply of agricultural goods but of the landscape itself in both its physical and social dimensions (see Chapter One).

Within a context of strong competition between direct producers furthered by the integrative trends operating across global agriculture, there is mounting pressure to continually intensify biological productivity by simplifying, standardising and speeding up production. This has imposed an increasingly industrial metabolism upon farming that rests on fossil fuel-driven technologies and the systematic manipulation of the genetic programming of plants and animals. For plant-based agriculture, modern farming techniques have transformed agrarian environments by cultivating extensive monocultures aimed at creating economies of scale through the specialised and concentrated production of a narrow range of crop species (Altieri 1995). These monocultures are enabled through the systematic application of synthetic fertilizer, the control of pests and weeds through chemical inputs, the compulsory provision of irrigation and the ongoing manipulation of plant and animal genomes, either through crossbreeding or genetic manipulation, to create hybrids with specific traits (Gliessman 2007).[1] Van der Ploeg perhaps overstates the point when he claims that farming in this mould is "reduced to a mere conversion of commodities (that might originate from anywhere) into other commodities (that might be destined for any location)" (2010: 4). However, he nonetheless captures the sense to which agriculture is increasingly disembedded from localised socio-ecological processes and is fundamentally transformed in an attempt to support ever-increasing aggregate yields. The purpose of this is to make agricultural processes and their underlying dynamics more amenable to human timescales of production, investment, exchange and accumulation (Boyd, Prudham and Schurman 2001).

Within the modernisation narrative of agrarian transformation, these trends are overwhelmingly presented as a natural evolutionary process driven by the creeping rationalisation of agricultural production through the ongoing spatial diffusion of efficient technologies and institutional forms (World Bank 2007). The resulting achievements of modern farming are highlighted as far-reaching, socially progressive and deeply necessary. Measured in terms of yields, the systematic reworking of agricultural practices has resulted in a consistently rising productivity that underscores the abundance of food at an aggregate level. Over the past 50 years, for example, the area devoted to agriculture has grown by around 12 per cent, whereas overall agricultural production has expanded by more than two and a half times (FAO 2011b). This consistent increase at a rate above that of population growth is attributed precisely to the application of modern agricultural technologies focused on the combination of improved varieties and breeds, specialised synthetic inputs, mechanisation and the controlled delivery of water in a continual drive to improve yields. Securing the provision of relatively cheap food, moreover, is concurrently seen as a foundational pillar of a more efficient division of labour that can support urbanisation, industrialisation and development both nationally and globally (Friedmann 1982).

There are, of course, readily acknowledged tensions with this petroleum-fuelled model of agricultural production. Intensive monocropping removes farming metabolisms from a wider network of biological renewal processes that hinge upon the networked interactions between plants and the wider ecosystem. As a result, a constant and growing application of external inputs are needed to supply crops with nutrient recycling, microclimate control, hydrological regulation, pollination and pest control (Altieri 1995; Altieri and Roge 2011). This raises the related issues of land degradation, toxicity, declining water resources, energy inefficiency and greenhouse gas emissions that are seen to be significant externalities and growing constraints upon modern agriculture. Notably, while yield level increases have remained above those of population, they have tended to taper off over the past three decades while rates of increase in synthetic input usage have surged (Bruinsma 2003). This suggests that the ever-heightening intensity of petroleum-driven farming – with the above-noted consequences for environmental health and climate change – is having less impact upon yield growth and possibly portends an escalating exhaustion of the paradigm (FAO 2011a; UNCTAD 2013; UNEP 2014). At the same time, such industrial farming is manifestly energy inefficient and a major driver of greenhouse gas emissions. According to some accounts, the industrial food system expends 10–15 energy calories to produce one calorie of food, constituting a reversal of the original purpose of agriculture (Lin et al. 2011; Martínez-Alier 2011).

Alongside concerns about the potential exhaustion of industrial farming, the growing productivity of agriculture has readily co-existed with the inability of a significant proportion of the world's population to gain adequate access to nutritious food. According to the International Fund for Agricultural Development, some 70 per cent of the developing world's 1.4 billion extremely poor people inhabit rural areas and frequently experience malnutrition despite their role within food production (World Bank 2007; IFAD 2010). In a world of relative abundance, there remains a large rural underclass that is beset with profound food insecurity and material poverty. It is precisely this population, moreover, that is considered as particularly vulnerable to climate change (IFAD 2013a).[2]

Within the framework of agrarian modernisation, the solution to such problems is twofold. First, increasing investment in biotechnological research is seen as necessary to produce a new generation of yield-enhancing technologies (Von Braun 2007; Baulcombe et al. 2009; Beddington 2011). Second, existing yields must be improved by accelerating the diffusion of existing techniques and technologies more deeply into the agrarian landscapes of the developing world. As encapsulated in the World Bank's broad yet strident 2008 World Development Report, this is to be achieved by integrating smallholders into global agro-food commodity chains so as to deepen the commercialisation of agricultural production, facilitate more market opportunities for producers and allow them to better access investment and technology. Smallholders are a particular focus of this strategy because they comprise the majority of the world's farmers, occupy 60 per cent of arable land and produce more than 80 per cent of the food consumed in the developing world (IFAD 2013b). Moreover, smallholder farming is seen as the primary barrier to solving the 'yield gap', which refers to the difference between the potential yield in an area under ideal conditions and technologies and the actual yield obtained. While the large differences in crop yields between and within countries can in part be attributed to differing agro-ecological conditions, it is the projection of substandard yields owing to differences in crop management practices, including the 'sub-optimal' use of inputs and farming techniques, that concerns policymakers (Bruinsma 2003). These traits are often seen as the direct result of an insufficiently modernized smallholder sector, making it a repeated target for development interventions.

From this perspective, it is the failure of agrarian modernisation to become sufficiently generalised that accounts for inadequate productivity and residual rural poverty. The task facing developing countries is, therefore, to reform the institutional parameters under which agricultural production takes place so as to facilitate a closer integration of smallholders into globalised agro-food chains as a vector of modernisation in the rural spaces of the global South. The projected expansion of market opportunities is envisaged to set in motion a diffusion of practices and technologies that, together, will create a more efficient smallholder sector orientated towards commercial production. As competitive forces and expanded opportunities lead to an increase in entrepreneurialism among smallholders, the resulting diffusion of new technologies are anticipated to help close the yield gaps between smallholders and elite farmers, increasing both productivity and incomes (Van Tran 2002; Toriyama, Heong and Hardy 2005). The ensuing efficiency gains stemming from increased agricultural productivity, moreover, are seen to facilitate a broader rationalisation of rural regions by allowing a more finely graded division of labour in which off-farm employment can expand to absorb the release of labour from unproductive smallholder farms.

For the World Bank, the primary policy questions stemming from this framework centre upon how to put suitable institutional structures in place to allow smallholders to access both the technological advances and the market opportunities that they will need to prosper. While the private sector is envisaged to drive the efficient organisation of value chains to increase market opportunities, it is the state that "corrects market failures, regulates competition, and engages strategically in public-private partnerships to promote competitiveness in the agribusiness sector and support the inclusion of smallholders and rural workers" (World Bank 2007: 8). This perspective fits into a broader neoinstitutionalist vision of development in which modernisation can only be achieved through overcoming the stultifying effects of traditional social institutions and forms of production that limit the rationalisation of agriculture (Taylor 2010). New institutions that can set appropriate incentive structures, therefore, are required to stimulate a more efficient agrarian sphere. A combination of price liberalisation, more clearly defined property rights and financial extension are anticipated to furnish the foundations for such an environment. Once appropriate institutional frameworks are established and increase aggregate efficiency, countries are anticipated to move forward over a typology of stages – agricultural, transitional, and urbanised – in a linear progression of development. The latter reflects a familiar modernisation script that is uncomfortably embossed over the heterogeneity of agrarian social structures present within and across countries. The telos of the World Bank's vision of agrarian futures is clear: "agriculture as a business, driven by entrepreneurship and vibrant markets, linked to a burgeoning urban economy" (Scoones 2009: 14).

A suitably entrepreneurial smallholder sector, therefore, is heralded as a fundamental pillar of the future global agricultural system. Smallholders can play such a role, however, only to the extent that they adequately adopt a suitably entrepreneurial orientation, integrate with broader commodity chains and embrace technological fixes to improve their productivity. As the World Bank notes, most smallholders will not be in such a position and this will inevitably create a process of social differentiation in which uncompetitive smallholders are rendered non-viable under fiercely competitive institutional parameters. The report, therefore, anticipates a major shift of labouring bodies out of agriculture through the abandonment of farming for either non-farm rural employment or migration to the urban realm. As a result, a principal regulatory task set out for governments is precisely to help rural populations whose labour is surplus to the requirements of a more efficient agricultural sector manage their transitions out of agriculture (Li 2009: 629). For most current smallholders, therefore, the vision of agrarian modernisation projects their future as one outside of farming, in waged labour and petty commodity production through an accentuation of rural-urban migration and the de-agrarianisation of rural spaces (Rigg, Salamanca and Parnwell 2012; Akram-Lodhi 2013).

Climate change and the antinomies of modernisation

The spectre of climate change is inserted into this vision not as a challenge to its embedded assumptions but as a confirmation of its existing biases. This is perhaps not surprising. In 2008, the same year as the publication of the World Bank's report, rapid price rises for general foodstuffs created a deep subsistence crisis for the world's urban and rural poor.[3] Notwithstanding the call of World Bank President Robert Zoellick for a 'new deal' on food, a familiar dish was hastily prepared by a combination of G8 countries and international institutions to reaffirm the centrality of agricultural modernisation through a corporate agenda of liberalisation and a second 'green revolution' founded upon biotechnology and transgenic crops.[4] This occurred despite little evidence to suggest that a problem in agricultural productivity was behind the crisis or that biotechnological solutions had any immediate role to play in resolving price hikes in the short or medium term (Stone and Glover 2011). Furthermore, it appeared to overlook a number of substantive irrationalities in the global food system that were revealed by a crisis in which agro-corporations enjoyed record profits while the global poor faced deepening food insecurity and increasing amounts of agricultural produce simultaneously were diverted to non-food purposes (Grain 2008). Despite this complex conjuncture of causes, including the growing role of financial speculation in shaping commodity prices, the response of leading international institutions was a emphatic yet familiar call to resolve such problems through increased productivity, with the Food and Agricultural Organization calling for a 70 per cent increase in food production by 2050 to stabilise market volatility and high prices (FAO 2011a).

The climate question has been similarly used to further reinforce these discursive parameters. Advocates of agrarian modernisation increasingly emphasise the importance of new seeds and inputs, driven by biotechnologies, to create plants that are more resilient to both adverse climatic conditions alongside biotic challenges such as pests and diseases that may proliferate in a warming world (Baulcombe et al. 2009; McIntire et al. 2009; Beddington 2011). While climatic change did not initiate this longstanding desire to produce new agrarian natures, it has certainly provided a rallying point around which biotechnologies have been justified (Patel 2013). In setting out a 'New Vision' for agriculture, for example, the 2010 World Economic Forum exemplified this trend. This report on the future of agriculture under climate change and price volatility projected the need to harness agriculture to drive a trio of goals, including food security, environmental sustainability and economic opportunity. It did so in familiar ways, citing the need for an innovation-driven agriculture wherein new agro-technologies implemented by local entrepreneurs can solve the multiple challenges that climate change poses:

> Building this pillar of the New Vision will require improvements across the supply chain to close yield gaps, promote efficient distribution, minimize waste and improve food access … It will also require technological breakthroughs to help farmers adapt to the consequences of climate change, enable production and mitigate risk under increasingly difficult conditions.
>
> (World Economic Forum 2010: 12–13)

This discursive emphasis on technology-driven productivity gains as the solution for the problems of the global food system obscures a series of core tensions from which two immediate points stand out. First, while farming is judged according to the efficiency of its productivity, there is no such reflection on the efficiency of contemporary consumption trends. A significant portion of the world's smallholder farmers are to be sacrificed on the alter of increasing farm productivity to service a global division of consumption that is starkly stratified, desperately inefficient in its use of biomass and energy and that propagates diets that – while helpful for embedded forms of accumulation – are nutritionally derelict (Patel 2007; Winson 2013). The modernisation narrative appears unable to question the consumption patterns it serves. As a recent UN Conference on Trade and Development report puts it, "[t]he current demand trends for biofuels, excessively meat-based diets and post-harvest food waste are accepted as given, rather than challenging their rationale" (UNCTAD 2013: 7). The idea of producing more represents an anxious desire to avoid having to face troubling questions around sharply uneven distributions of production and consumption through which access to food is highly skewed in a global level. Much as the original Green Revolution was seen as a way to accelerate agricultural growth without necessitating fundamental changes to rural power structures, a biotechnological revolution implicitly offers a means to avoid facing the social barriers to food access within and across nations (Patel 2013).

Second, the push for agricultural modernisation is done in the name of market efficiency at the same time as it constructs institutional frameworks that privilege the oligarchic power of corporate agribusiness. The concentration present at the corporate end of agro-food commodity chains is pronounced, with not only the centralisation of capital within agro-input producers, marketers and retailers, but new interlinkages between such corporate entities and financial entities that readily speculate on changing prices (Isakson 2014). This centralisation and concentration of capital within agriculture ranges from the agro-input conglomerates that presently have oligopolistic hold upon the production and marketing of seeds and synthetic inputs, through to the consolidated power of supermarkets in shaping agro-food commodity chains from the distribution end (Weis 2007; Howard 2009). The United Nations Environment Programme, for example, notes how by 2005 the largest ten seed corporations controlled 50 per cent of all commercial seed sales, the top five grain trading companies controlled 75 per cent of the market and the largest ten pesticide manufacturers supplied 84 per cent of all pesticides (UNEP 2014: 12).

As a result, the discourse of increasing opportunities that is embedded within the modernisation narrative singularly fails to analyse the characteristics of a food regime dominated by global agro-food corporations in which monopoly power across an integrated food system means that food and agricultural markets are "predatory, constraining and indeed forcing the choices of emerging capitalist farmers" (Akram-Lodhi 2008: 1160). These trends include the privatisation of knowledge through intellectual property rights which, as the International Assessment of Agricultural Knowledge, Science and Technology for Development notes, operates within a regulatory structure that benefits patent holders – typically corporate product manufacturers – rather than the rural communities that have developed genetic resources over millennia (IAASTD 2009). Indeed, a model of privatised knowledge-driven change translates directly into the strongly projected technological role for climate proofing agriculture, which has opened a new profit frontier for agro-industry through the patenting of 'climate ready' genes by the agro-industrial seed conglomerates (McMichael 2009).

Adaptation meets the agrarian question

By situating the question of adaptation within a broader terrain of unequal power structures at a global level, we can better come to terms with the political dimensions of what it means to read contemporary change through the adaptation–modernisation nexus within specific agrarian environments. In this regime of representation, climate change is seen as hastening the workings of longstanding processes of agrarian change in which rural regions must become more fully integrated into global divisions of production and consumption. The project of adaptation must simultaneously further this process of mopping up the historical residues of an incomplete transition to agrarian modernity while mediating its most brazen contradictions.

The discourse of modernising rural regions through select governmental interventions, however, is a longstanding narrative rooted in the political economy of capitalist transformation and subsequently reshaped through the colonial encounter (Cowen and Shenton 1996; Li 2007). For colonial bureaucrats in nineteenth century South Asia, for example, the doctrines of classical political economy pointed the way towards a more 'rational' usage of land, labour and water that required a concerted programme of planned engineering that could rewrite both the social and physical landscape in ways that could reform 'wasteful' practices (Gidwani 2008). These technologies were part of what summoned the agrarian into being as a site of governmental regulation that interlaced the administration of agricultural production with the movement of agricultural commodities across vast territories and the concomitant management of surplus extraction. As David Ludden puts it, agrarian history first appeared as a chronicle of state policy whose impact "was measured in the endless dance of numbers on agrarian taxation, rent, debt, cropping, output, living standards, technology, demography, land holding, contracts, marketing and other money matters" (1999: 8). Indeed, the essence of the modernist project, both then and now, is to disable 'non-modern' forms of life by dismantling their facilitating conditions and putting in place new conditions to produce governing effects to re-arrange the conditions of agrarian life in a way that obliges subjects to transform themselves in a deliberate and 'improving' direction (Mitchell 1990; Asad 1992; Scott 1999).

While this 'will to improve' has legitimised itself through a perpetual re-ordering of agrarian environments in order to harness their productive potentials, it has simultaneously and irrevocably been forced to mitigate the inevitable dislocations that ensue (Li 2007). The establishment of agrarian capitalism has never been a creeping process of gradual rationalisation through the evolution of increasingly efficient institutional forms. Rather, it has required prolonged, violent and far-reaching transformations tied to the commodification of land and labour and driven by processes of forceful social engineering often undertaken directly through coordinated state power (Cowen and Shenton 1996). The inherently conflictual processes that such trends have unleashed repeatedly and consistently involved the fusion of expanding commercial integration with historically embedded forms of power. As Gyan Prakash notes, in the South Asia context the rule of commodities and markets took shape in and profited from structures ranging from peasant production to plantation slavery, even though it represented them as its opposite (1997: 22). It is therefore is necessary to conceptualise the ways in which the formal rationality of agrarian rationalisation has been inextricably entangled with processes that distribute both the risks and rewards of agricultural production in vastly uneven ways. The story of increasing agrarian productivity is simultaneously one of dispossession, accumulation and vulnerabilisation within the ongoing transformation of agrarian environments (Bernstein 2010; Akram-Lodhi 2013; McMichael 2013).

For agrarian political economists, these concerns are often framed under the rubric of the 'agrarian question' that, as Karl Kautsky wrote at the turn of the twentieth century, concerns "whether, and how, capital is seizing hold of agriculture, revolutionising it, making old forms of production and property untenable and creating the necessity for new ones" (1988: 12). The notion of the agrarian question has become a primary analytical passageway into the contemporary transformation of agrarian social relations (Akram-Lodhi and Kay 2010a, 2010b). There is, of course, no singular form of agrarian transformation under these complex dynamics. The question nonetheless raises a set of crosscutting concerns about the way in which rural populations experience pronounced social differentiation in which the full or partial proletarianisation of rural labour occurs in complex forms and new forms of surplus extraction emerge through a range of means, including waged labour, debt relations, land rentals, monopolised input markets and coerced labour service. Within the framework of the agrarian question, such forces create a pervasive fragmentation of rural social classes. This polarisation occurs not as a failure of modernity to arrive but as a consequence of its inherently uneven tendencies. Willem van Schendel (2006) is emphatic on this point when he argues that, despite 150 years of large-scale capitalist production within agriculture, South Asia has refused to bear out the assumption that the agrarian capitalism would produce a clear separation of capital and labour patterned upon a linear transformation in a modernist mould.

On the contrary, while the expansion of wage labour during and following the colonial period has been considerable, the resulting agrarian social order refuses to replicate any simple model of transition. For much of South Asia, a dynamic and finely graded diversity of labouring relations has emerged in which much of the rural population has been rendered insecure as agricultural producers yet face manifold constraints to becoming waged labourers in either the rural or urban realms. Many households rely upon members operating as transitory labour – 'wage hunters and gatherers' in Jan Breman's (1994) striking term – moving in patterns of circular migration between agricultural activities, labouring as rural wage labourers and temporary employment in urban informal sector. They may also, however, stubbornly persist in keeping a foothold in agriculture, either for subsistence or commercial purposes, by deploying household labour often at low levels of productivity. Such a bulwark may be viewed as a partial preserve against the vulnerability implicit within a commodified agrarian environment in which social reproduction is dependent upon fickle and often coercive forces (van der Ploeg 2010). This active recomposition of agrarian environments has frequently led to a situation in which rural households depend – often tenuously – on a diversity of social relationships and forms of labour, from waged work, to market-orientated petty commerce to subsistence agriculture on small plots (Shah and Harriss-White 2011).

There is no reason to romanticise this proclivity to strategically moderate vulnerabilities within austere agrarian environments. In the provocative words of Russian agrarian economist Vasily Chayanov, "one cannot fail to recognise that in the course of the most ferocious economic struggle for existence, the one who knows how to starve is the one who is best adapted" (1991: 90). What such an emphasis does do, however, is raise the complex politics of land and vulnerability in agrarian environments undergoing complex transitions within which climatic change is one active element. As Haroon Akram-Lodhi and Cristobal Kay put it, processes of de-peasantisation, semi-proletarianisation, re-peasantisation and petty commodity production under capitalist dynamics cannot be seen as aspects of a linear process. Rather, they form "dynamic and recurrent manifestations of multifaceted and contradictorily changing patterns of social and economic relations that continually and complexly reconfigure rural labour regimes, and hence the agrarian question" (Akram-Lodhi and Kay 2010b: 280). As such, the bare elements of the South Asian experience mapped above cannot be seen as a universal paradigm of contemporary agrarian change. Indeed, as the following chapters show, there are notable divergences in agrarian trajectories across South Asia which demonstrate incredible heterogeneity between regions owing to differing historical processes and agro-ecological characteristics (Banaji 2002). That said, it is useful to pick up on three common elements that emerge as key tensions within agrarian environments in the chapters that follow: the question of labour, issues of power and risk within smallholder production and the question of ecological rifts.

Labour

A determined aspect of the adaptation–modernisation nexus highlighted above is its emphasis upon a necessary transition of livelihoods away from agricultural pursuits according to both economic rationality and an adaptation imperative in the face of climatic change. As the following chapters note, however, an exit from smallholder farming is rarely a choice made by rational calculating households that simply weigh up livelihood options based upon a simple projection of varied opportunities. On the contrary, smallholders are often reluctant to sever their ties to the land and will often seek to invest further familial labour or incur growing debts to maintain agricultural production on small plots. In the World Bank's terms, such reluctance to move out of agriculture reflects a "deep inertia in people's occupational transformation" that is ultimately a blockage to modernisation (World Bank 2007: 26). This position reflects a longstanding modernist trope that the innate cultural values of peasants are a barrier to efficient rural divisions of labour (Hoselitz 1952; see Taylor 2010). As the following chapters show, far from being irrationally attached to agriculture owing to an enduring cultural atavism, the rural poor are – out of necessity – shrewd in the ways they seek to secure their social reproduction and assert dignity in the face of intractable burdens. Indeed, for Gidwani and Sivaramakrishnan, the rural poor in much of South Asia actively construct livelihoods that mediate the many risks they face, including forms of cosmopolitanism through circular migrations between different physical and social spaces in which they are able to "deploy the technologies of one to some advantage in the other" (2004: 345).

It is precisely on the basis of such 'survival entrepreneurialism' that some neo-institutional theorists have been attracted to the informal economy as a latent reservoir of repressed entrepreneurs, forced underground by the arduous regulations of the state (de Soto 1989). There is no need, however, to romanticise the desperately hard choices and strategies made by these social classes. For those forced out of agriculture, they must seek alternative employment on fickle labour markets with few assets, skills or networks. In moving out of agriculture on these adverse terms, their subsequent level of disempowerment makes them subject to conditions of labour that reinforce their vulnerability and poverty. As Jan Breman puts it:

> To the extent that these many hundreds of millions are incorporated into the production process it is as informal labour, characterized by casualized and fluctuating employment and piece-rates, whether working at home, in sweatshops, or on their own account in the open air; and in the absence of any contractual or labour rights, or collective organization. In a haphazard fashion, still little understood, work of this nature has come to predominate within the global labour force at large.
>
> (Breman 2009: 29)

It should be emphasised, moreover, that such informal work is heavily segmented along gender lines with women disproportionately present in the informal sector and specifically in those jobs that are most insecure and poorly remunerative in both rural and urban realms (Chant and Pedwell 2008). Such injustice is reflective of the broader gendered division of entitlements, property and responsibilities that permeates many agrarian environments and that structures labour across the household, field and off-farm economies in ways that create durable power relations and inequalities between men and women (FAO 2011c).

The fragility of employment opportunities within heavily segmented informal labour markets underscores the reluctance of many rural households to relinquish fully their hold of land as a means to preserve some degree of autonomy against such conditions of vulnerability (van der Ploeg 2008; Ferguson 2013). Maintaining a direct hold – however tenuously – over the means of social reproduction can facilitate what Sanyal terms a 'need economy', an ensemble of economic activities undertaken for the purpose of meeting direct needs, as distinct from activities driven by an imperative towards systemic accumulation (Sanyal 2007: 209). It is the rightful fear that, under full commodification of land and labour, livelihoods are subject to external dynamics and power relations over which households have little form of control. In short, tempering their vulnerability to such forces can provide a strong motivation for smallholder agriculturists to maintain at least a partial hold on land (Shah and Harriss-White 2011). The World Bank appears at times to recognise such issues, noting that for those unable to generate sufficient assets and skills to manage a successful transition out of agriculture, "only social protection can ease their poverty" (World Bank, 2007: 202). This seemingly essential point – made even more vital in the context of climatic change – is nonetheless submerged under the discussion of enabling markets, new technologies and moving populations out of agriculture (Li 2009). As the following chapters make clear, we should be profoundly cautious before we wager the future of those ejected from agriculture on the fanciful idea that informal labour markets can provide a means towards secure livelihoods, particularly as climate change impacts are projected to fall most heavily on the peri-urban slum dwellers precisely because of their socio-economic and physical marginalisation (UN-Habitat 2011).

Power and risk

Within the parameters of the adaptation–modernisation nexus, agrarian transformation is represented through a seductive imagery of positive-sum games in which new opportunities arise for smallholders to avail themselves of new market opportunities and technologies. Establishing the institutional conditions to enlarge the scope of such opportunities, therefore, represents a win-win situation in which farmers become more competitive, productive and resilient and consumers gain access to a wider and cheaper array of products. Simultaneously, even those displaced through the consolidation of a more efficient smallholder sector are ultimately the beneficiaries of a more rational division of labour in which urban and off-farm employment provides a better set of livelihood options (Oya 2009).

For those smallholders that seek to remain – even partially – in agriculture, farming has its own risks and associated power relations. As noted above and detailed in the following chapters, however, the institutional framework imposed upon smallholders is not simply one designed to increase opportunities but a specific means of imposing new constraints and compulsions (Akram-Lodhi 2013). In the face of intense cost-pressures, smallholders consistently encounter stern pressures to adopt commercial farming through input-induced intensification as a means to expand yields and maintain household incomes. To do so, smallholders are increasingly drawn into markets for key inputs – seeds, land, water, electricity, fertiliser, pesticides, credit and retail – over which they exert little control. These inputs must be sourced through commercial networks outside the farm and are often accessed on the basis of extended credit, which draws smallholders into new forms of dependency and surplus extraction. The international peasant movement Via Campesina captures this new dynamic of capital formation in which direct control of land is no longer the lynchpin of capital accumulation within the agrarian environment. Although land control is still important, engrained structures of power are often constructed primarily through networks that "control loans, materials supply, the dissemination of new technologies, such as transgenic products, on the one hand, and those that control national and international product warehousing systems, transportation, distribution and retail sales to the consumer, on the other hand" (Via Campesina, cited in McMichael 2012: 684).

These power relations, it should be emphasised, are not merely the result of 'institutional imperfections' or 'market failures', as described in the euphemistic terms of neoinstitutionalism (Amanor 2009). Rather, they represent specific accumulation strategies in which unequal concentrations of market power are used as a form of surplus extraction and the displacement of risk (see Chapter Four). They represent the outcome of historically unequal distributions of core assets and frequently are bolstered through the greater political influence that comes with concentrated command over such productive resources. In the modernisation narrative, contract farming offers a way for smallholders to tap directly into a more lucrative relationship with larger scale agro-food enterprises. This often is represented as a form of beneficial incorporation into wider networks to access opportunities from which smallholders – and particularly women farmers – were previously excluded. Through such incorporation, smallholders are envisaged to plough new and mutually beneficial routes into global agro-commodity chains that offer better returns for a more diverse range of products (World Bank 2007).

What tends to be lost this representation, however, is how contract farming establishes relations between parties that hold strongly unequal degrees of power to shape the distribution of risks and rewards within the relationship (Singh 2012; Akram-Lodhi 2013). As Muhammad Jan and Barbara Harriss-White note, the contractual relationship between producers and large corporations turns smallholders into effectively wage-labour status. The dominant partner shapes the type of crop, choice of inputs, the amount of credit and timing of sales (Jan and Harriss-White 2012). For Jan van der Ploeg, this socio-ecological shift results in a notable transformation of power relations within agricultural production. Land that once served as a form of landed power or peasant autonomy increasingly has become a link in longer chains that tie smallholders into exogenous and frequently more powerful interests and projects (van der Ploeg 2010: 4).

By its very nature, contract farming requires a strong standardisation of products, detailed quality controls and a finely graded scheduling of crop production. In short, it presumes an industrialised farming model and the associated inputs, technologies and social relations to standardise, speed up and intensify production. This is why contract farming historically has been associated with larger farms that operated to economies of scale with large production volumes and better access to credit, water, technology, labour and informational networks (Singh 2012). Large farms have built upon highly inequitable distributions of such assets precisely to consolidate their position within rural hierarchies in a self-reinforcing manner. To the extent that smallholders can move into contract farming within agro-food commodity chains, their scattered nature, small production volumes and historical marginalisation from stable sources of irrigation and credit amplify the risks involved. Undoubtedly, the potential rewards of contract farming can be considerable, as contracts for high-value products can aid investment to produce increased yields and improve incomes. The resulting risks, however, are notable. Farmers often incur significant debts in order to purchase the necessary inputs to facilitate industrial farming, and this creates new forms of dependency and surplus extraction with financial operatives while increasing the catastrophic risk of harvest failure (see Chapter Seven). It is this enduring fragility of smallholder agriculture leveraged on debt and tied to input intensive production that accentuates the threats to the underlying socio-ecology posed by climatic change (Altieri and Koohafkan 2008).

As this makes clear, the relations that structure agrarian environments are not simply predicated upon surplus extraction but simultaneously incorporate a vastly unequal distribution of risks. Part of the problem is that the deepening subsumption of agriculture within agricultural commodity chains creates the grounds upon which external technological solutions appear to be the only ones possible. For example, as regional biodiversity decreases in the wake of input-intensive monocultures, a series of biological checks upon pest populations are removed, reinforcing a strong dependency upon commodified chemical inputs and the commercial networks between corporations, merchants and farmers that facilitate their sale. The work of Glenn Stone on the generalisation of biotech cotton in the semi-arid region of Telangana, India, exemplifies this trend. Stone charts how the industrialisation of agricultural practices in the Warrangal region created a progressive de-skilling of smallholders who became increasingly dependent upon forms of knowledge and inputs over which they have very little control. A longstanding facet of the Green Revolution model, this trend became entrenched in Warrangal during in the 1990s when, under the liberalisation of agriculture, smallholders increasingly adopted a form of cotton monocropping as a response to severe cost pressures on agriculture. Through the reduction of biological barriers to the evolution and spread of pest species, smallholders encountered a cycle of escalating pesticide use that proved increasingly ineffective against rampant bollworm infestations. As Stone terms it, "never before had they relied so heavily on a crop with so many pests, so little natural resistance, and in such apparent need of sprays" (Stone 2011b: 393).

It was in this socio-ecological context that biotech cotton emerged in the mid-2000s as a projected solution to a problem caused by the earlier round of industrial agriculture. For farmers, any technologies that promised to offset the onslaught of bollworms seemed a necessary and welcome entrant into input markets. The deepening dependency upon external technology to counteract field-level problems, however, raises important questions around vulnerability in an era of climatic change because such technologies profoundly rework the balance of power within the agrarian environment. In drawing agriculture into the ambit of the laboratory, farmers have become embedded at the bottom rung of a highly commodified form of agricultural practice whose implementation rests on bodies of knowledge from which they are systematically excluded (Stone 2011a). While corporate agricultural input producers tend to represent their goods as 'solutions' to pressing agricultural issues, the kinds of issues they address are not fixable problems but moving targets. As Boyd, Prudham and Schurman note, "efforts to further control and subordinate biological systems to the dictates of industrial production will almost inevitably generate new risks and vulnerabilities for the production process, not to mention unforeseen externalities" (2001: 562). In this manner, there emerges a fallacy of composition associated with the generalised movement of smallholders onto biotechnologies. As more and more farmers in a given region turn to a particular form of cropping – using a narrow range of seeds, herbicides and pesticides – the quicker the spread of pests that are immune to such technologies. Several years into the biotech era, bollworms had indeed receded only to be replaced with a major aphid problem. Subsequently, the incipient development of resistance in pink bollworm to biotech cotton is an issue of growing concern.

For agro-industry, the active flux of agrarian environments are simply part of their accumulation strategy, generating a constant flow of problems that need new solutions through commodified inputs from which they extract substantial flows of surplus. Stone notes how, for biotech cotton in Telangana, agrarian technologies were increasingly shaped by the commercial dictates of the turnover time of corporate capital, which needed new brands on a yearly basis. For smallholders, this pace of technological change and the resulting generation of new challenges in the field from using these technologies occurs at a pace that prohibits an effective process of social learning. This excludes them from becoming active participants in managing production processes. As a result of this estrangement and dependency, seed-purchasing trends were driven not by rational choices based on field conditions but by corporate marketing strategies and localised fads (Stone, Flachs and Diepenbrock 2013). In Philip McMichael's terms, the displacement of former agro-ecological farming practices and knowledge generation with commoditised agro-inputs facilitates a 'real subsumption' of smallholders to capital. It reduces them to quasi-waged labourers and soothes the appropriation of agricultural resources along the value-chain as profits for processors, retailers, agro-dealers and traders (McMichael 2013). This opens up not only new accumulation strategies but also a new vector of vulnerability predicated precisely upon the technologies to which agrarian modernisation sees as the saving grace for smallholders. At a time when climate change is altering the socio-ecological parameters of agricultural production, smallholders are increasingly alienated from knowledge and control over the active production of their lived environments.

Metabolic rifts and climatic change

It is useful to note how this emphasis on the dynamic interplay between farmers, pests, synthetic inputs and genetically modified crops points to lived environments as active sites of socio-ecological production. As the biotech example highlights, while humans have undoubtedly proved to be prolific agents of environmental production, we are nonetheless inescapably drawn into a complex field of relationships with varied elements of the non-human world that fundamentally shape the dynamics of agrarian environments. In this respect, it is necessary to ask where the ecology is hidden away within the agrarian question. It is notable that within the 'agrarian question', the drivers of rural change have tended to be represented as emphatically anthropocentric, and the social categories of capital and labour seemingly impose themselves upon the natural substrate of the rural landscape. In their comprehensive review of the field, for example, Haroon Akram-Lodhi and Cristobal Kay (2010b: 269) posit ecology as one of seven 'dimensions' of the contemporary agrarian question. In so doing, they advocate examining the ways in which 'ecological relationships' impinge upon agrarian social relations. While their framework is both sophisticated and encompassing, it nonetheless reinforces a conceptualisation of the agrarian in anthropocentric terms within which ecological considerations may (or may not) set limits or exercise external impacts. This kind of framework has tended to reinforce the idea of nature as a tapestry upon which humanity designs its own, internally contested, plans. Nature becomes a vital resource that can be marshalled as part of social relations through forms of possession and exclusion.[5]

As established in Chapter One, however, relationships of production and consumption, abundance and scarcity and security and vulnerability are made in and through relationships in which the boundaries of nature and society are intrinsically blurred. Our environments do not present simple constraints upon human actions. Rather, they provide the essential metabolic forces upon which human agency is facilitated. For some working within political economy traditions, these socio-ecological relations can best be captured by the notion of a 'metabolic rift' (Foster 1999; Wittmann 2009; Schneider and McMichael 2010). Building on Marx's discussion of a collapse of nutrient recycling between countryside and city within agrarian capitalism, this concept has been generalised to capture a systematic rupture in the metabolic relation between humans and nature under capitalism. It emphasises the essential contradiction between an industrialised agriculture driven by capitalist value relations and the ecological foundations upon which labouring activities are based (Foster, Clark and York 2010). Approaching the issue in this manner does have some utility. It draws attention towards processes that might lead to a tipping point: a moment of crisis in socio-ecological relations in which the particular relationships that produce a lived environment take a radically different turn under the weight of their own contradictions. Although sympathetic to the need to bring metabolism into the agrarian question, however, the idea of the metabolic rift appears to represent a more radical version of resilience theory and suffers from maintaining a similar conceptual framework predicated open the idea of mutually influencing yet ontologically separate social and natural systems (see Chapter Four). Like Jason Moore, I am concerned that the metabolic rift framework tends to assign arbitrary boundaries that separate off the 'social' and the 'ecological' dynamics that are seen to pull apart, reaching some sort of absolute point of contradiction, namely 'the rift' (Moore 2011).

In focusing on the metabolic rift as a moment of rupture between society and nature, the approach marginalises how produced socio-ecologies are both productive and destructive of particular forms of life. This returns us to Blaikie and Brookfield's point that "one person's degradation is another person's accumulation" (1987: 14). It is, therefore, more useful to consider the question raised by the metabolic rift concept in terms of a conjuncture of socio-ecological processes working across scales that generates tension, strain, struggle and constant transformation. In this way, the utility of the metabolic rift notion can be found not in terms of an absolute point of rupture in which intrinsically 'social' and 'ecological' processes fracture but rather as an analysis of the inherent tendency of capitalist socio-ecologies to create strong and mutual tensions between the valuation of commodities and the processes that underpin their production, circulation and the dispersal of associated waste products. This is to focus on the scalar processes that link the transformations of lived environments across both time and space (Mitchell 2012).

On this score, we must conceptualise the futures of agrarian environments in terms of the ongoing socio-ecological shifts in which biophysical processes and social energies are interweaved, co-productive, uneven and tension-laden. Climatic change enters such landscapes not as an exogenous factor but as a shift in the socio-ecological parameters by which livelihoods are formed. Across agrarian Asia, for example, the rapid but uneven melt of Himalayan glaciers, influenced by anthropogenic processes including greenhouse emissions and more localised black carbon pollutants, will be a significant driver of socio-ecological change. Himalayan glaciers feed seven of Asia's major rivers – the Ganga, Indus, Brahmaputra, Salween, Mekong, Yangtze and Huang Ho – therein sourcing irrigation waters for over a billion people (Pomeranz 2009; Xu et al. 2009). In the Ganga river system alone, the loss of glacier melt water is projected by the IPCC to reduce July-September flows by two-thirds, causing water shortages for 500 million people and 37 per cent of India's irrigated land. On this basis, the IPCC projects that India will reach a state of water stress before 2025 when the availability of water falls below 1,000 m^3 per capita (Cruz et al. 2007: 484). This is not to suggest that such biophysical shifts pose determining social constraints. Water stress already is experienced by large segments of the rural populations in these areas based on questions of access rather than absolute quantities of water in a given setting (see Chapters Six and Seven). Rather, it is to highlight that the shifting socio-ecological parameters of agrarian environments are objects of ongoing social struggles from above and from below that seek to actively reshape the production of risks and rewards. For some, the outcomes of conflicts over changed flows of water may offer new accumulation strategies. For others, they may spell the end of agrarian livelihoods, therein pushing a further flow of bodies out of rural regions into the expanding urban slums. This is a question that requires historical analysis and I seek to map out precisely these tensions in the following three chapters.

Notes

1 Although not a focus of this chapter, modern agriculture also is defined by the factory farming of animals, in which meat production is organised upon industrial techniques of mass production (Emel and Neo 2011).

2 Cheap imports of specific foods also are key to containing the tensions inherent to growing inequalities of income and the expansion of urbanised, gendered and racialised

underclasses in Western countries. In particular, a food regime based on cheap yet nutritionally derelict industrial food knits into a broader expansion of precarious low-wage work supplemented by informal coping strategies including corporate-sponsored food banks in the United States, a phenomenon that Ryan Foster has termed 'divide and sponsor' (Foster 2008; see also, Soederberg 2014).

3 Josette Sheeran, the head of the UN's World Food Programme, described the impact of rising food prices as a process of deepening immiseration: "For the middle classes, it means cutting out medical care. For those on $2 a day, it means cutting out meat and taking the children out of school. For those on $1 a day, it means cutting out meat and vegetables and eating only cereals. And for those on 50 cents a day, it means total disaster." Cited in 'The New Face of Hunger', *The Economist,* April 17, 2008.

4 G8 Statement on Global Food Security, Hokaido, Japan, July 8, 2008. http://www.g8summit.go.jp/eng/doc/doc080709_04_en.html.

5 Part of this latent anthropocentrism is perhaps given by the desire to combat neo-Malthusian projections of intrinsic environmental limits that are seen to strongly determine the warp and weft of social transformation. In circumventing a singular logic of environmental scarcities that impose themselves in a direct manner on forms of social organising and the incidence of conflict, political ecologist have tended to emphasis how localised power structures, connected to global geo-politics of resource utilisation, themselves construct the scarcities that neo-Malthusians then impute to be 'natural' (Peluso and Watts 2001).

References

Agrawal, A., and N. Perrin. 2008. "Climate Adaptation, Local Institutions, and Rural Livelihoods." *IFRI Working Paper* (# W08I-6): 1–17.

Akram-Lodhi, A. H. 2008. "(Re)imagining agrarian relations? The World Development Report 2008: Agricultural for Development." *Development and Change* 39(6): 1145–61.

Akram-Lodhi, A. H. 2013. *Hungry for Change: Farmers, Food Justice and the Agrarian Question*. Halifax, NS: Fernwood.

Akram-Lodhi, A. H., and C. Kay. 2009. *Peasants and globalization: Political Economy, Rural Transformation and the Agrarian Question*. London; New York: Routledge.

Akram-Lodhi, A. H., and C. Kay. 2010a. "Surveying the agrarian question (part 1): Unearthing foundations, exploring diversity." *Journal of Peasant Studies* 37(1): 177–202.

Akram-Lodhi, A. H., and C. Kay. 2010b. "Surveying the agrarian question (part 2): Current debates and beyond." *Journal of Peasant Studies* 37(2): 255–84.

Altieri, M. A. 1995. *Agroecology: The Science of Sustainable Agriculture*. Boulder, CO: Westview Press.

Altieri, M., and P. Koohafkan. 2008. *Enduring Farms: Climate Change, Smallholders and Traditional Farming Communities*. Penang: Third World Network.

Altieri, M., and P. Roge. 2011. "The Ecological Role and Enhancement of Biodiversity in Agroecosystems." Pp. 15–32 in *Agriculture, Biodiversity and Markets: Livelihoods and Agroecology in Comparative Perspective*, edited by Stewart Lockie and David Carpenter. London: Earthscan.

Amanor, K. 2009. "Global food chains, African smallholders and World Bank governance." *Journal of Agrarian Change* 9(2): 247–62.

Asad, T. 1992. "Conscripts of Western Civilisation " Pp. 333–51 in *Dialectical Anthropology: Essays in Honor of Stanley Diamond*, edited by C. Gailey. Gainesville, FL: University Presses of Florida.

Banaji, J. 2002. "The metamorphoses of agrarian capitalism." *Journal of Agrarian Change* 2(1): 96–119.

Baulcombe, D., I. Crute, B. Davies, J. Dunwell, M. Gale, J. Jones, J. Pretty, and W. Sutherland. 2009. *Reaping the Benefits: Science and the Sustainable Intensification of Global Agriculture*. London: The Royal Society.

Beddington, J. 2011. *Foresight: The Future of Food and Farming*. London: The Government Office for Science.

Bernstein, H. 2010. *Class Dynamics of Agrarian Change*. Halifax, NS: Fernwood.

Blaikie, P. M., and H. C. Brookfield. 1987. *Land Degradation and Society*. London: Methuen.

Boyd, W., W. S. Prudham, and R. Schurman. 2001. "Industrial dynamics and the problem of nature." *Society & Natural Resources* 14(7): 555–70.

Breman, J. 1994. *Wage Hunters and Gatherers: Search for Work in the Urban and Rural Economy of South Gujarat*. Delhi; New York: Oxford University Press.

Breman, J. 2009. "Myth of the global safety net." *New Left Review* 59: 29–37.

Bruinsma, J. (Ed.). 2003. *World Agriculture Towards 2015/2030: An FAO Perspective*. London: Earthscan.

Chant, S., and C. Pedwell. 2008. *Women, Gender and the Informal Economy: An Assessment of ILO Research and Suggested Ways Forward*. Geneva: ILO.

Chayanov, A. V. 1991. *The Theory of Peasant Co-Operatives*. London: B. Tauris.

Clapp, J. 2012. *Food*. Cambridge, UK; Malden, MA: Polity.

Cowen, M., and R. Shenton. 1996. *Doctrines of Development*. London: Routledge.

Cruz, R., H. Harasawa, M. Lal, S. Wu, Y. Anokhin, B. Punsalmaa, Y. Honda, M. Jafari, C. Li, N. Huu Ninh, C. Cambridge, and University Press. 2007. "Asia – Climate Change 2007: Impacts, Adaptation and Vulnerability." In *Contribution of Working Group II to the Fourth Assessment Report of the Intergovernmental Panel on Climate Change*, edited by M. L. Parry, O. F. Canziani, J.P. Palutikof, P. J. van der Linden, and C. E. Hanson. Cambridge: Cambridge University Press.

de Soto, H. 1989. *The Other Path: The Invisible Revolution in the Third World*. London: I.B. Taurus.

Emel, J., and H. Neo. 2011. "Killing for Profit: Global Livestock Industries and their Socio-Ecological Implications." Pp. 67–83 in *Global Political Ecology*, edited by Richard Peet, Paul Robbins, and Michael Watts. London: Routledge.

FAO. 2011a. "The State of Food Insecurity in the world: How Does International Price Volatility Affect Domestic Economies and Food Security?", edited by Food and Agriculture Organization. Rome: FAO.

FAO. 2011b. *The State of the World's Land and Water Resources for Food and Agriculture: Managing Systems at Risk*. Rome: FAO.

FAO. 2011c. *Women in Agriculture: Closing the Gender Gap for Development*. Rome: FAO.

Ferguson, J. 2013. "How to do things with land: A distributive perspective on rural livelihoods in Southern Africa." *Journal of Agrarian Change* 13(1): 166–74.

Foster, J. B. 1999. "Marx's theory of metabolic rift: Classical foundations for environmental sociology." *American Journal of Sociology* 105(2): 366–405.

Foster, J. B., B. Clark, and R. York. 2010. *The Ecological Rift: Capitalism's War on the Earth*. New York: Monthly Review Press.

Foster, R. 2008. "Corporate Social Responsibility and Labour Market Discipline: Contesting Social Reproduction in Low-Wage America." Pp. 117–36 in *Global Economy Contested: Power and Conflict across the International Division of Labour*, edited by Marcus Taylor. London: Routledge.

Friedmann, H. 1982. "The political economy of food: The rise and fall of the postwar international food order." *American Journal of Sociology* 88(S): 248–86.

Gidwani, V., and K. Sivaramakrishnan. 2004. "Circular Migration and Rural Cosmopolitanism in India." Pp. 339–67 in *Migration, Modernity and Social Transformation in South Asia*, edited by Filippo Osella and Katy Gardner. New Delhi: Sage.

Gidwani, V. K. 2008. *Capital, Interrupted: Agrarian Development and the Politics of Work in India*. Minneapolis: University of Minnesota Press.

Gliessman, S. R. 2007. *Agroecology: The Ecology of Sustainable Food Systems*. Boca Raton, FL: CRC Press.

Goodman, D., and M. Watts. 1997. *Globalising Food: Agrarian Questions and Global Restructuring*. London; New York: Routledge.

Grain. 2008. "Making a Killing from Hunger." Pp. 1–6 in *Against the Grain*. Barcelona: GRAIN.

Hoselitz, B. 1952. "Non-economic barriers to economic development." *Economic Development and Cultural Change* 1(8): 8–22.

Howard, P. 2009. "Visualizing consolidation in the global seed industry: 1996–2008." *Sustainability* 1(4): 1266–87.

IAASTD. 2009. *Agriculture at a Crossroads*. Washington, DC: IAASTD.

IFAD. 2010. *Rural Poverty Report 2011*. Rome: IFAD.

IFAD. 2013a. *The Adaptation Advantage: The Economic Benefits of Preparing Small-Scale Farmers for Climate Change*. Rome: IFAD.

IFAD. 2013b. *Smallholders, Food Security and Agricultural Development*. Geneva: IFAD/UNEP.

Isakson, S. R. 2014. "Food and finance: The financial transformation of agro-food supply chains." *Journal of Peasant Studies* 41(5): 754–82.

Jan, M. A., and B. Harriss-White. 2012. "The three roles of agricultural markets: A review of ideas about agricultural commodity markets in India." *Economic and Political Weekly* 47(52): 39–53.

Kautsky, K. 1988. *The Agrarian Question: In Two Volumes*. London: Zwan.

Li, T. 2007. *The Will to Improve: Governmentality, Development, and the Practice of Politics*. Durham, NC: Duke University Press.

Li, T. 2009. "Exit from agriculture: a step forward or a step backward for the rural poor?" *Journal of Peasant Studies* 36(3): 629–36.

Lin, B., M. Jahi Chappell, J. Vandermeer, et al. 2011. "Effects of industrial agriculture on climate change and the mitigation potential of small-scale agro-ecological farms." *CAB Reviews: Perspectives in Agriculture, Veterinary Science, Nutrition and Natural Resources* 6(20): 1–18.

Liu, A. 2010. "The birth of a noble tea country: On the Geography of colonial capital and the origins of Indian tea." *Journal of Historical Sociology* 23(1): 73–101.

Ludden, D. 1999. *An Agrarian History of South Asia*. Cambridge: Cambridge University Press.

Martínez-Alier, J. 2011. "The EROI of agriculture and its use by the Via Campesina." *Journal of Peasant Studies* 38(1): 145–60.

McIntire, B., H. Herren, J. Wakhungu, and R. Watson. 2009. *Agriculture at a Crossroads: International Assessment of Agricultural Knowledge, Science and Technology for Development*. Washington, DC: Island Press.

McMichael, P. 2009. "Contradictions in the global development project: geo-politics, global ecology and the 'development climate'." *Third World Quarterly* 30(1): 247–62.

McMichael, P. 2012. "The land grab and corporate food regime restructuring." *Journal of Peasant Studies* 39(3–4): 681–701.

McMichael, P. 2013. *Food Regimes and Agrarian Questions*. Halifax, NS: Fernwood.

Mitchell, D. 2012. *They Saved the Crops: Labor, Landscape, and the Struggle over Industrial Farming in Bracero-Era California*. Athens, GA: University of Georgia Press.
Mitchell, T. 1990. "Metaphors of power." *Theory and Society* 19(5): 545–77.
Moore, J. 2011. "Transcending the metabolic rift: a theory of crises in the capitalist world-ecology." *Journal of Peasant Studies* 38(1): 1–46.
Oya, C. 2009. "The World Development Report 2008: Inconsistencies, silences and the myth of 'win-win' scenarios." *Journal of Peasant Studies* 36(3): 593–601.
Patel, R. 2013. "The long green revolution." *Journal of Peasant Studies* 40(1): 1–63.
Patel, R. C. 2007. *Stuffed and Starved: Markets, Choice and the Battle for the World's Food System*. Toronto: HarperCollins.
Peluso, N. L., and M. Watts. 2001. "Violent Environments." Pp. 3–37 in *Violent Environments*, edited by Michael Watts and Nancy Lee Peluso. Ithaca, NY: Cornell University Press.
Pomeranz, K. 2009. "The great Himalayan watershed: Agrarian crisis, mega-dams and the environment." *New Left Review* 58: 5–40.
Prakash, G. 1997. "Colonialism, Capitalism and the Discourse of Freedom." Pp. 9–25 in *Peripheral Labour? Studies in the History of Partial Proletarianization*, edited by S. Amin and M. van der Linden. Cambridge: Cambridge University Press.
Rigg, J., A. Salamanca, and M. Parnwell. 2012. "Joining the dots of agrarian change in Asia: A 25 year view from Thailand." *World Development* 40(7): 1469–81.
Sanyal, K. 2007. *Rethinking Capitalist Development: Primitive Accumulation, Governmentality, and Post-Colonial Capitalism*. New Delhi: Routledge.
Schneider, M., and P. McMichael. 2010. "Deepening, and repairing, the metabolic rift." *Journal of Peasant Studies* 37(3): 461–84.
Scoones, I. 2009. "Livelihoods perspectives and rural development." *Journal of Peasant Studies* 36(1): 171–96
Scott, D. 1999. *Refashioning Futures: Criticism after Postcoloniality*. Princeton, NJ: Princeton University Press.
Shah, A., and B. Harriss-White. 2011. "Resurrecting scholarship on agrarian transformations." *Economic & Political Weekly* XLVI(39): 13–20.
Singh, S. 2012. "New markets for smallholders in India: Exclusion, policy and mechanisms." *Economic & Political Weekly* 29(52): 95–106.
Soederberg, S. 2014. *Debtfare States and the Poverty Industry: Money, Discipline and the Surplus Population*. London: Routledge.
Stone, G. D. 2011a. "Contradictions in the last mile: Suicide, culture and e-agriculture in rural India." *Science, Technology and Human Values* 36(6): 759–90.
Stone, G. D. 2011b. "Field versus farm in Warangal: Bt cotton, higher yields and larger questions." *World Development* 39(3): 387–98.
Stone, G. D., A. Flachs, and C. Diepenbrock. 2013. "Rhythms of the herd: Long term dynamics in seed choice by Indian farmers." *Technology in Society* 36(1): 26–38.
Stone, G. D., and D. Glover. 2011. "Genetically modified crops and the 'food crisis': Discourse and material impacts." *Development in Practice* 21(4–5): 509–16.
Taylor, M. 2010. "Conscripts of Competitiveness: Culture, institutions and capital in contemporary development." *Third World Quarterly* 31(4): 561–79.
Toriyama, K., K. Heong, and B. Hardy (Eds.). 2005. *Rice Is Life: Scientific Perspectives for the 21st Century*. Manila: IRRI.
UN-Habitat. 2011. *Cities and Climate Change: Initial Lessons from UN-Habitat*. Geneva: United Nations.
UNCTAD. 2013. *Trade and Environment Review 2013*. Geneva: United Nations.

UNEP. 2014. *Assessing Global Land Use: Balancing Consumption With Sustainable Supply*. Geneva: United Nations Environment Programme.

van der Ploeg, J. D. 2008. *The New Peasantries: Struggles for Autonomy and Sustainability in an Era of Empire and Globalization*. London: Earthscan.

van der Ploeg, J. D. 2010. "The peasantries of the twenty-first century: The commoditisation debate revisited." *Journal of Peasant Studies* 37(1): 1–30.

van der Ploeg, J. D. 2013. *Peasants and the Art of Farming: A Chayanovian Manifesto*. Halifax, NS: Fernwood.

van Schendel, W. 2006. "Stretching Labour Historiography: Pointers from South Asia." Pp. 229–61 in *Coolies, Capital and Colonialism: Studies in Indian Labour History*, edited by Rana Partap Behal and Marcel van der Linden. Cambridge: Cambridge University Press.

Van Tran, D. (Ed.). 2002. *Sustainable Rice Production for Food Security*. Rome: FAO.

Von Braun, J. 2007. *The World Food Situation: New Driving Forces and Required Actions*. Washington, DC: International Food Policy Research Institute.

Weis, A. J. 2007. *The Global Food Economy: The Battle for the Future of Farming*. London: Zed Books.

Winson, A. 2013. *The Industrial Diet: The Degradation of Food and the Struggle for Healthy Eating*. Vancouver: UBC Press.

Wittmann, H. 2009. "Reworking the metabolic rift: La Via Campesina, agrarian citizenship, and food sovereignty." *Journal of Peasant Studies* 36(4): 805–26.

World Bank. 1982. *World Development Report 1982*. Oxford: Oxford University Press.

World Bank. 2007. *World Development Report 2008: Agriculture for Development*. Oxford: Oxford University Press.

World Bank. 2009. *World Development Report 2009: Development and Climate Change*. Washington, DC: World Bank.

World Economic Forum. 2010. *Realizing a New Vision for Agriculture: A Roadmap for Stakeholders*. Geneva: WEF.

Xu, J., R. E. Grumbine, A. Shrestha, M. Eriksson, X. Yang, Y. Wang, and A. Wilkes. 2009. "The melting Himalayas: Cascading effects of climate change on water, biodiversity, and livelihoods." *Conservation Biology* 23(3): 520–30.

6 Pakistan – historicising 'adaptation' in the Indus watershed

The devastating monsoon floods that struck Pakistan in 2010 swamped almost one-fifth of the country and caused significant loss of life and considerable damage to crops and physical infrastructure. Alongside a number of concurrent 'erratic weather events', ranging from heat wave-induced forest fires in the Russian north to mudslides in western China, numerous observers drew a causal link between the Pakistani floods and anthropocentric climate change. Scientists at the World Meteorological Office, for example, specified a direct relationship between anthropocentric climate change, higher than normal Atlantic Ocean temperatures and the intense nature of the monsoon rains over the western Himalaya (Gronewold 2010). Pakistani officials also highlighted these linkages. Foreign minister Shah Mehmood Qureshi stated forcibly at the United Nations General Assembly in 2010 that the post-flood situation reconfirmed Pakistan's extreme vulnerability to the adverse impacts of climate change. Such claims were supported by new indices of climate change vulnerability, such as the Maple Croft index produced at Columbia University, which placed Pakistan as the foremost country at risk from the adverse impacts of climate change in 2010.

Within Pakistani government circles, this proclamation of extreme risk to climate change represented a notable discursive shift. During the 2000s, the potential threat of climate change had largely been sidelined under the weighty presence of enduring security concerns related to the military dictatorship, the Afghan war and the spectre of internal terrorism. The 2010–11 floods, however, catalysed the mainstreaming of climate change into the policy discourse of the Pakistani People's Party government of Asif Zardari. The sudden renaming of the Ministry of National Disaster Management as the Ministry of Climate Change reflected this shift. It instantly created an institutional body charged with incorporating climate change concerns into a broader policy strategy to promote sustainable development. From this new institutional configuration, policy documents were swiftly forthcoming. A National Climate Change Policy released in 2011 presented a first framework aimed at addressing climate change-induced threats to water, food and energy security, as well as minimising the impact of natural disasters on human life, health and property (Government of Pakistan 2011). The latter was subsequently expanded upon within the context of a national sustainable development strategy drawn up by the new Ministry of Climate Change in 2012 (Government of Pakistan 2012).

For Pakistan, the creation of a multifaceted climate change strategy aimed at reducing vulnerability to short-term hazards and long-term environmental change seemed both necessary and, in comparison to neighbouring countries, overdue. Yet despite this new urgency, the manner in which the Pakistani government is approaching the issue raises a number of important questions related to the concerns raised in the first half of this book. In posing adaptation in terms of a series of technical and managerial solutions to an externally generated threat, the Pakistani government's approach silences the power relations that are inherent to the production of vastly unequal agrarian environments in the Indus watershed. Ultimately, for the Pakistani government, the essence of climate change adaptation is the pursuit of change so that things can remain the same. As a consequence, the narrative of adaptation serves to legitimise policy strategies that exist in great tension with their stated goals of sustainability and equity.

In contrast, this chapter seeks to resituate the production of vulnerability within agrarian Pakistan by historicising climate change as part of a longer trajectory of producing agrarian environments. It does so by situating contemporary shifts within the context of agrarian transformations that have unfolded over the past century and a half. Two key points are made. First, in contrast to the notion that Pakistan is facing a novel and externally generated threat, the chapter emphasises how the Indus region comprises an intensely manufactured environment in which the vulnerability of rural populations to drought and floods remains intimately tied to the socio-ecological engineering of the Indus basin ongoing since colonial times. The social dynamics of agrarian change are therein inseparably intertwined with active ecological transformations that have created environments that are simultaneously productive and destructive for their human protagonists. Second, this historical reshaping of the region's socio-hydrology is interlaced with the highly uneven control of land, water and labour in a context of changing integration with regional and world markets for agricultural produce. Notwithstanding important regional variations, the outcome of these processes has been a highly stratified rural social order characterised by concentrated land ownership, pervasive landlessness, interlocked markets and the presence of bonded labour.

To be clear, in charting this history of socio-ecological production in Pakistan, the chapter does not offer a comprehensive historical account of the Indus region as a whole, which is beyond the scope of this work. Nor does it provide an exhaustive account of the insecurities faced by people in the wake of the recent floods. Rather, it seeks to take seriously Greg Bankoff's claim that disasters are processes in which hazardous events "represent moments of catharsis along a continuum whose origins lie buried in the past and whose outcomes extend into the future" (2001: 30). It does so as a means to substantiate the claim raised earlier in the book that we need to bring into clear relief the different forms of power that produce agrarian environments as a prerequisite for understanding how climatic change becomes embedded within a wide array of social practices and biophysical processes. This is undertaken as a way to better conceptualise the various meanings, possibilities and the contradictions of climate change adaptation as a contemporary form of political practice and to suggest ways we can think beyond the idea of adaptation.

I elaborate this argument in four sections. The first provides an overview of the climate change adaptation strategy established by the Pakistani government, emphasising the way it constructs climate change as an exogenous environmental threat to the nation as a whole. Section two then begins to historicise climate change adaptation by charting the impact of colonial experience upon land ownership and water usage under changing climatic conditions. The colonial period fundamentally reshaped the relationship between social classes in the agrarian sphere by privatising land and integrating agricultural production into colonial trading circuits. The third section focuses on the campaign to reform the hydrological infrastructure of the country through an internationally funded programme of canalisation and Green Revolution technologies in the postcolonial period. While such socio-technological engineering increased agricultural productivity, the reshaping of Pakistan's agrarian relations was predicated upon inequitable control over land and water that strongly conditioned the relative vulnerability of the agrarian population to climate variability. The fourth section continues the analysis to the present, emphasising how the fault lines inherent to this historically formed agrarian environment have been laid bare within the post-floods agrarian environment. Here, control over key productive assets (land, water, credit) is central to the relative ability of different sections of the rural population to respond to the floods. In the conclusion, the chapter reflects briefly upon the policy implications of the analysis. While land reform has been placed on the backburner of Pakistani development strategy, this is a political elision that is reinforced by the discursive parameters of climate change adaptation. In contrast, the chapter argues that land tenure issues should be central to any meaningful debate on the futures of agrarian environments.

Conceptualising climate change adaptation in Pakistan

In a country facing manifold political, economic and security challenges, the issue of climate change received relatively little attention at the Pakistani governmental level over the first decade of the new millennium.[1] As noted above, this changed dramatically with the 2010 and 2011 floods, which created a new awareness of the threat posed by increased climate variability and reaffirmed the lack of governmental preparedness for hazards of this magnitude. A rapid discursive realignment took place, and the Pakistani government now argues that anthropogenic climate change will cause a number of hydroclimatic shifts that have critical implications for economic growth and social stability throughout the country (Government of Pakistan 2012). These include further floods, accentuated droughts, increased thermal stress on crops, advancing seawater intrusion into the Indus delta and the erosion of fragile ecosystems across the national territory. In a region characterised by a significant dependence on agriculture for employment, high levels of rural poverty and a public infrastructure that is badly overburdened, all of these impacts are projected to threaten agricultural- and natural resource-based livelihoods. To contextualise the gravity of these projections, it should be noted that agriculture contributes about 21 per cent to the GDP and generates employment for 45 per cent of the country's labour force, while 60 per cent of the rural population depend directly on this sector for their livelihood (Ministry of Finance 2012). As such, the effects of climatic change on household welfare, particularly in rural areas, are envisaged to be significant and largely detrimental (ICUN, Ministry of Environment and DIFD 2011).

In response, the 2011 National Climate Change Policy presents itself as a comprehensive framework for the development of a climate change action plan (Government of Pakistan 2011). Furthermore, the 2012 Sustainable Development Strategy incorporated these adaptation measures into a broader policy synthesis of economic and social policy aims (Government of Pakistan 2012). Following the argument set out in Malik Aslam Khan's (2011) National Economic and Environmental Development Study produced by the Ministry of Environment, these documents emphasise that Pakistan is a "prime victim of global 'climate injustice' that will bear the burden of the impacts with a minimal contribution to this global problem" (Government of Pakistan 2012: 49). Such a claim is indeed well established, as highlighted by the IPCC and the Stern Review among others (Cruz et al. 2007; Stern 2007). The report then proceeds to argue that the country's extreme vulnerability to this externally generated threat is a feature of its "geographic location, elevation as well as demographics" (Government of Pakistan 2012: 50). As it elaborates, most of Pakistan is composed of extremely arid regions in which water for agriculture, industry and consumption is drawn from the Indus river system that flows downwards from rain, snow and glacier melt in the western Himalaya during the spring and summer. While the impacts of climate change upon these complex hydroclimatic processes are still debated (Archer et al. 2010), the threats of glacial melt caused by rising temperatures alongside the amplified likelihood of extreme rainfall are suggested to exacerbate water shortages and increase floods throughout the Indus basin. These threats, moreover, are argued to be accentuated by demographic expansion wherein the inexorable growth of population amplifies demands upon limited and already stressed water resources (Government of Pakistan 2012: 48). In short, the underlying causal factors of vulnerability to adverse impacts of climatic change in Pakistan are reduced to the accidents of geography coupled to a Malthusian invocation of natural population pressures.

On this basis of this understanding of the causality of climate change, the *National Climate Change Policy* frames the process of climate change adaptation in terms of an increasingly scientific management of resources and threats. Although neither document explicitly sets out what it means by adaptation, they implicitly follow the IPCC approach discussed in Chapter Three. The latter, strongly influenced by the resilience paradigm, describes adaptation as a process of fortifying the ability of societies to adjust to the effects of changing climatic phenomena so as to alleviate adverse effects in a way that can maintain livelihoods and safeguard or ameliorate material wellbeing and social identity (IPCC 2001). Vulnerability here is accorded not to specific social groups or individuals but either to the unit of the country as a whole or to specific economic sectors, such as the vulnerability of agriculture to climate change (ICUN, Ministry of Environment and DIFD 2011). The overriding concern is to strengthen the existing parameters of Pakistani society in the face of an externally generated environmental challenge. Climate change adaptation, therein, is presented in terms of an increasingly scientific management of resources in which institutional reforms and new technologies are the necessary means to curtail vulnerability and secure sustainability.

Accordingly, the policy establishes eight sectors in which adaptation concerns need to be mainstreamed into existing development strategies. For the first seven of these – Water Resources, Agriculture and Livestock, Human Health, Forestry, Biodiversity, Vulnerable Eco-Systems and Disaster Preparedness – the strategy is predicated upon a series of technical and managerial reforms to diminish the adverse impacts of climate variability. For example, there is a strong emphasis on upgrading water storage and management, transitioning to drought resistant crops, better regulating land use to curtail environmental degradation and on improving flood prevention and preparedness. The eighth and final category – Socio-Economic Factors – addresses the social context for adaptation. In typical adaptation speak, it stipulates that the poor are more vulnerable to climate change effects owing to high dependence on natural resources, their limited technological capacity and insufficient financial resources to cope with climatic extremes. To address this vulnerability, the document highlights further measures comprising the need for curbing population growth, ensuring the awareness and preparedness of local communities and promoting new technologies of crop production and pest management (Government of Pakistan 2011: 20–23).

To be clear, technical and managerial strategies are essential elements of any comprehensive approach to dealing with the impacts of climatic change. The physical strengthening and lining of existing canals and waterways alongside improving efficiency of water use across rural and urban Pakistan, for example, are ongoing concerns and must be placed high upon the governmental agenda as Pakistan increasingly moves towards a condition of water scarcity (see Briscoe, Qamar and Contijoch 2005; Kugelman and Hathaway 2009). Notwithstanding the importance of these issues, however, they cannot be adequately addressed in a manner that is singularly devoid of social and historical context. In the case of water stress, as discussed below, numerous authors have demonstrated how scarcity in Pakistan is socially constructed through power relations that unevenly structure the control and usage of irrigation water within rural society (see, for example, Akram-Lodhi 2001a; Mustafa 2002). On this basis, neither the accidents of geography nor the spectre of a burgeoning population can be understood in abstraction from the socio-ecological relations through which both resources and risks are inequitably distributed across Pakistan. To invoke Timothy Mitchell (2002), what appears as nature – the inexorable rise of population in land of geographic and hydrological extremes – is already shaped by forms of power, technology, expertise and privilege that can be unpacked to reveal a very different picture of powerless, inequality and exploitation.

In what follows, I address this point by drawing on the perspective set out in the first part of this book to demonstrate significant fractures in the ideological foundations and political practices inherent to the Pakistani government's approach to adaptation. Demographic pressures and water stress are important factors, yet they derive their significance not from natural constraints that condition human interactions with nature but from the historical processes that have driven the socio-ecological transformation of the Indus watershed over the past 150 years. The latter have consolidated a profoundly stratified agrarian environment in which vulnerability to various expressions of climatic change is endemic yet is produced relationally and concentrated among specific social classes and genders. On this basis, such vulnerability can only be understood in terms of the hierarchically ordered socio-ecology of rural Pakistan wherein the relative security of particular social groups is achieved through socio-ecological dynamics that produce the relative insecurity of others. To set climate change within these parameters is to indicate the profoundly reactionary underpinnings of technical and managerial approaches and to challenge the meanings and practices of adaptation in the Pakistani context.

Colonisation, canalisation and climate variability

While the Indus region has a rich agrarian history that traces back some five millennia, within its modern period the advent of British colonialism is recognised to be a critical period of institutional change that comprehensively reshaped the socio-ecology of the region.[2] After annexing Sindh and the Punjab during the 1840s, the British were eager to simplify and taxonomise the structure of land tenure in order to more effectively exercise rule and extract revenues. To do so, they created a system of formalised land rights that were superimposed over existing land tenure relations. At the same time, they imposed taxes on a fixed cash rate according to the estimated value of the land. While these efforts encountered varying forms of accommodation and resistance from local populations, several core impacts emerged over the following decades.

Notably, by assigning formal property titles to land, the colonial state turned landlords, who previously enjoyed socially mediated claims to a share of the produce of the land, into formalised landowners who now possessed the land as a private and alienable resource. This transformation of land rights made an indelible impact upon the region's agrarian social relations. In the Punjab, the *mahalwari* land settlement policies served to consolidate the position of the dominant peasant caste that had risen to prominence during the collapse of the Mughal authority in the preceding century (Gazdar 2011: 15–20). The new landlord class soon formed the bedrock of colonial policy: receiving title, recognition and revenue-sharing arrangements from colonial administrators and also becoming a mainstay of the British Raj's military forces following the 1857 revolt (Ali 2002).

In the province of Sindh, the British attempted to impose a *ryotwari* system of land tenure, in which taxation revenues would be paid directly by the cultivator to the colonial state. In theory, this arrangement might have benefited the local cultivating *hari* caste. However, the ability of the British to govern was heavily dependent on the cooperation of the existing *wadero* landlords (Cheesman 1981). To reinforce this power base, the British acceded to political expediency and proclaimed the *waderos* to be both owners and cultivators of the land, leaving the *hari* designated essentially as serfs. Not only did this designation transform and consolidate the power base of the *waderos*, it engrained a new and enduring agrarian structure across the province that has often been described as 'feudal' given the concentrated hold on economic and political power it afforded. As such, far from simply building on pre-existing social structures, British colonialism thereby transformed them. Colonialism was therein deeply engaged in a process of what David Washbrook describes as the creation and perpetuation of what would subsequently be labelled 'traditional' social structure across South Asia (Washbrook 1993).

These new agrarian social relations underwent further transition from the 1880s onwards when the colonial government initiated the engineering of the Indus watershed to open up new areas for irrigated agriculture. There were two primary motivations. First, irrigation was an essential part of expanding the cropped area so as to increase exports of raw materials into the British industrialisation drive and improve tax revenues. Second, centrally governed irrigation provided a method of consolidating state power over agrarian life, as small independent water systems were integrated into the centralised control system (Gilmartin 1988). As Daniel Haines has argued, the development of Sindh's irrigation system was predicated upon increasing state intervention in agrarian life to render a 'disordered' society more easily governable (Haines 2011; see also Mustafa 2001). From the 1880s onwards, successive rounds of canalisation reshaped the socio-ecology of both the Punjab and Sindh, expanding settled agriculture at the direct expense of nomadic pastoralism. Under the justification of promoting a loyal and agriculturally progressive landowner class, the settlement policies for canal colonies tended to preserve the existing monopoly of the landlords by excluding the landless and poor peasants from grants of land (Hamid 1982). In both provinces, the agrarian elite benefited enormously from monopolisation of new ecological and agricultural opportunities (Ali 2002).

These new structures of landownership, taxation and access to water played a substantial role in distributing opportunities and risks across the agrarian environment. As local agricultural production was increasingly incorporated into colonial markets, a new commercial logic attended the expansion of irrigation, as exemplified in a major expansion of cash cropping for both national and international markets. During this period, Karachi was transformed from a provincial backwater into the British Raj's fourth largest seaport, serving as an export centre for wheat, cotton and oilseeds drawn from the interior of Punjab and Sindh. This new nexus of agricultural production, irrigation expansion and commercialisation fed into the rise of a network of merchant/money-lenders drawn primarily, though not exclusively, from Hindu castes (Cheesman 1982; Mishra 1982). The latter were able to use the extension of credit to connect the localised socio-ecologies of irrigated agricultural production across the Indus basin to the commodity circuits that fed the ongoing metabolisms of British industrialisation.

This colonial boom, however, had decidedly polarising repercussions. Among the small peasantry, colonial policies that insisted on taxation based on the value of the land instead of a portion of the produced crop facilitated an endemic vulnerability to commercial and ecological shifts. Falls in international crop prices or the occurrence of drought obliged smallholders to turn to moneylenders to make up for shortfalls. Given that land was now both fully alienable and rapidly increasing in price, moneylenders tended to make loans against land mortgages. Leveraged by such debts, land began to further concentrate in the hands of powerful landlords and merchant/moneylender classes (Hamid 1982; Nazir 2000). In 1901, fearful that indebtedness might undermine political and social stability, the colonial authorities passed the Land Alienation Act, which reserved land ownership for specific castes that they deemed to be agrarian. While this curbed the power of professional merchant-moneylenders, it did not alleviate the vulnerability of the peasantry but instead tended to consolidate the power of large landowners who added the functions of moneylenders to their control over land and water resources (Mishra 1982: 41; Gazdar 2011).

Under these socio-ecological dynamics, the colonial period witnessed a substantial concentration of land and a correlate growth in the size of landless agrarian classes. By the 1930s, close to 48 per cent of the agricultural population in the Punjab was landless and, of the remaining landowning population, almost 60 per cent were poor peasants with landholdings under five acres. Concurrently, the landlord class that comprised less than four per cent of the agricultural population controlled around half the agricultural land in the province (Hamid 1982). Landlords and exporters were well positioned to make sizeable profits on the sale of wheat and other foodgrains upon colonial markets, whereas the peasantry and landless workers found themselves in a situation of profound vulnerability that translated into endemic food insecurity. Furthermore, landlords sought to monopolise more productive lands located in areas best served by the canal irrigation. The small peasantry, in contrast, were pushed further down the canal tailheads where water flow was erratic, therein escalating the impact of variations in rainfall, or onto land prone to flooding.

Two points are specifically worth highlighting in the context of climatic change. First, a pattern of locking in differential exposure to floods along class lines – so brutally demonstrated in 2010 – had already become established at the turn of the previous century. Significant portions of the small-holding peasantry were pushed onto marginal lands through the workings of debt and dispossession in a way that constructed an uneven human topography of vulnerability to flooding (see also, Mustafa 2005). Second, those small landholders pushed down the tailheads of canals were also more exposed to drought under the new irrigation systems. Notably, this shift in the socio-ecology of the Indus occurred parallel to a period of noted climatic change due to a major fluctuation in the El Niño

Southern Oscillation cycle (Davis 2002). The socially constructed vulnerability to climatic shifts was brutally exposed by a period of fickle monsoons that led to seven famines in the 40 years between 1861 and 1901. Most cruelly, the drought of 1878–79 claimed the lives of some 1.25 million people across the north-western area of the British Raj alone (Davis 2002: 51). As this makes clear, a deadly confluence of climatic change and constructed vulnerability is far from a novel feature of Pakistan's agrarian history.

The socio-ecology of vulnerability in postcolonial Pakistan

The human catastrophes of the late Victorian period forced a moderation of colonial rule, yet this did little to attenuate the concentration of land across the Indus watershed, which continued into the 1930s. While more benign colonial rule moderated the imposition of taxation upon local populations, it was not ultimately enough to save the British Raj. Independence and the creation of Pakistan in 1947, accompanied by the horrors of partition, placed new strains on rural regions that had been shaped by a century of colonialism. On the one hand, partition imposed a new political border across the Indus river system, separating upstream control of some tributaries from downstream users and creating a tense international politics of water usage (Wescoat, Halvorson and Mustafa 2000). On the other, the stressed migration of tens of thousands of people across the Punjab created new agrarian tensions (Gazdar 2011). The Muslim League, which rose to power in the newly formed state, had attempted to enlist the rural landless classes into the political programme for independence by promoting land reforms and other forms of economic improvement. At the same time, however, the party was closely knitted into the landlord classes that provided much of its political cadre (Niazi 2004). This created a politically charged environment from which the military dictatorship of Ayub Khan seized power in 1958 to consolidate the status quo. In an attempt to address the productive deficiencies of Pakistani agriculture and simultaneously allay the expectations of the rural underclasses, the regime promulgated an agrarian reform in 1959 that had been fervently debated over the best part of the decade. While it set out to achieve land redistribution, the final act was widely critiqued owing to the high limit set on individual holdings (500 acres per person) and the relative ease that landowners found in getting around the regulations, often by simply dividing estates between family members. As such, despite some efforts towards regulating landlord-tenant relations, it precipitated no fundamental change to existing power structures across the agrarian environment (Joshi 1974).

With land reform failing to address the substantial socio-economic contradictions of agrarian Pakistan, the military regime looked to alternative forms of mediating these tensions. In this regard, Akmal Hussain wryly noted that the introduction of Green Revolution technology in the 1960s allowed the Pakistani ruling classes to breathe a sigh of relief (Hussain 1988). Whereas under labour-intensive agriculture small plots using familiar or hired labour had a yield advantage, the new technology shifted the balance in favour of capital inputs and economies of scale. This facilitated an 'elite-farmer strategy' in which high-yield variety (HYV) seeds in combination with chemical fertilisers and pesticides facilitated significant yield gains. As Hussain concluded, the Green Revolution made it possible to accelerate agricultural growth without substantive changes to the rural power structure. The benefits of increased yields, it was expected, would energise the entire structure of the rural economy, providing trickle-down benefits to all social groups. As Mushtaq Gadi put it, the objective of the Green Revolution was "not only to engender the process of technological innovation but also to engineer socio-political relations" (2003: 274).

The expectations of a technologically driven social revolution, however, proved naively optimistic in the face of the inherited power structures of agrarian Pakistan (Niazi 2004). Large landholders were in a privileged position to monopolise the benefits of technological change owing to superior access to credit to purchase the new inputs, economies of scale through tractorisation and above all their superior access to irrigation. The replacement of labour through mechanisation and the resumption of land formally leased by landlords resulted in evictions and a proletarianisation of rural labour, which pushed up poverty rates. The latter were only tempered in the 1970s and 1980s through the remittances that returned to rural areas from migrants to both Pakistani cities and the Gulf. Agrarian struggles that resulted in a land reform in the 1970s under the populist government of Zulfikar Bhutto. Despite redistribution of some marginal lands, the latter did little to break the back of concentrated landed property, and a promised second round of reforms in 1977 was thwarted by a brutal coup d'état.

While land provided one of the ecological foundations of agrarian social structure, control over hydrological flows was equally critical to the deployment of Green Revolution technology. HYV seeds were heavily dependent on the measured application of water at specific times in the growing cycle and this meant that the Green Revolution relied on a new socio-ecology of water sourcing, distribution and usage. The means for this agrarian hydrology were provided primarily through an expansion in the scale and scope of canal irrigation allied to the proliferation of tubewell technology (Briscoe, Qamar and Contijoch 2005). In 1960, the Indus Waters Treaty between India and Pakistan provided the political basis for a new round of major engineering under the Indus Basin Development Programme. This programme built towards a more integrated and centrally controlled hydrological system that introduced major new dams and inter-river water transfers, therein creating one of the world's largest irrigation and hydropower systems (Wescoat, Halvorson and Mustafa 2000). Control over fickle flows of water stemming from the Himalayan watershed was projected as both an indicator of Pakistan's fate as an independent nation and a prerequisite for agrarian modernisation and food security through Green Revolution technology.

For Daanish Mustafa and David Wrathall, however, integrated river basin development created a Faustian bargain in which the potential gains from amplified control over the region's hydrology needed to be set against the production of new hazards, such as waterlogging, sedimentation build-up, increased salinity and flooding on a greater scale.[3] They argue that the pattern of large-scale hydrological engineering swapped high frequency, low-intensity flood events for low-frequency, high-intensity flood events. While the 2010–2011 floods are the most devastating examples of this, 20 major floods occurred in the 50 years following independence alone (Mustafa and Wrathall 2011). Simultaneously, by destabilising the water cycle, hydrological engineering profoundly imbalanced the socio-ecology of the Indus basin. By adding more water through irrigation than the natural drainage system could diffuse, groundwater levels around canal zones began to rise markedly, leading to widespread land degradation through waterlogging and salinity that still plagues the region (Gadi 2003; Mustafa 2010).

The primary strategy to confront waterlogging involved the use of tubewells, which appeared to serve two purposes simultaneously. Abstracting groundwater helped lower the water table, thereby tempering land degradation through waterlogging and salination. At the same time, tubewells provided a form of 'irrigation on demand' that was essential for the extension of HYV seed agriculture. While tubewells were initially promulgated through public installation, private construction by landowners soon became the norm. As a consequence, the benefits of tubewells were notably stratified along both geographic and class lines. On the one hand, wells remain heavily concentrated in the Punjab, where groundwater is fresh and able to be used as an agricultural input. In much of Sindh, however, the use of tubewells has proved more problematic as groundwater is brackish and unsuited for agricultural use. Here problems of salination remain acute and widespread, with 70 to 80 per cent of soils classified as moderately to severely salinized (Mustafa 2010: 9). On the other hand, tubewell ownership was and remains highly concentrated among larger landowners. The latter were able to afford their private installation or could access formal credit sources. Large landowners also were able to profit from economies of scale in combining irrigation and tractor use. As the first substantial wave of private tubewell installation took place over the 1970s and 1980s, 70 per cent were installed by famers owning more than 25 acres and only four per cent by farmers with fewer than 13 acres (Zaidi 2005: 30). Currently, should smallholders or tenants find away to access credit to drill wells, they face the growing over-exhaustion of groundwater resources. As Qureshi et al. (2010) note, falling water tables and increasing salt content entail that groundwater will become more expensive and inferior in quality, with notable consequences for agricultural productivity.

The class differentials surrounding tubewells mirrored those of canal water use. Rashida Haq (2007) finds a highly skewed and growing inequality of size of holding and access to irrigation, therein reinforcing the link between control of land and the benefits of access to water. In part, this is because large landholdings have been accumulated at the head of canals where flows of water are highest before decreasing in volume thereafter. A study in the Punjab conducted by Bhutta and van der Velde (1992), for example, highlighted how numerous canal offshoots remained dry for up to 90 per cent of the year despite their supposedly perennial nature. Concurrently, control over canal irrigation occurs through a mix of bureaucratic procedure – with local officials notably susceptible to the pressures of politically important landowners – and local institutions that are shaped by those same local power structures in which the need to placate landowners who control water sources reinforces the dependency of marginal cultivators and tenants (Khan 2007). As Haroon Akram-Lodhi succinctly elaborated through a detailed case study of irrigation projects in Mardan, water serves as a critical source of social power (Akram-Lodhi 2001a: 331). Unequal control over water therein reflects and serves to bolster a broad range of agrarian power relations that consolidate the inequities of rural Pakistan.

Debt relations are a further fulcrum of this power structure and a major means through which rural surpluses are appropriated and vulnerability created. Numerous studies have highlighted how a majority of small producer and tenant households in the Punjab and Sindh remain permanently in debt, drawing upon landlords or local merchants for cash advances necessary to purchase key agricultural inputs, from seeds to fertilizer, for the year ahead (Lieten and Breman 2002; Hussain 2003). As one recent study noted, interest rates tend to range from 35 per cent of the loan advanced to 100 per cent (CARE International 2012: 28). These cycles of debt, in which the amount is never fully cleared, force the farmer to sell their crop to the provider of the loan at a price and time that benefits the latter, a phenomenon known as 'interlocking markets' as discussed in Chapter Four (see Akram-Lodhi 2001b: for a seminal study in northern Pakistan). While the government offers minimum procurement prices for all main crops, smallholders and tenant farmers can receive rates well below these owing to relations of dependency through endemic indebtedness. The Pakistan Human Development Report of 2003, for example, noted how an overwhelming proportion of rice, cotton and wheat produce is sold to traders and landlords as intermediaries, which the authors estimate leads to lost income of about 20 per cent of the output value (Hussain 2003: 67).

These overlapping forms of power produce intense vulnerability within rural Pakistan along the lines of caste and class. In contemporary Sindh, for example, a pervasive lack of access to land forces many *hari* cultivators to accept heavily imbalanced sharecropping contracts from landlords that are further reinforced through ties of debt and access to canal irrigation. In addition to handing over a specified percentage of the crop, the tenant will also be required to undertake unpaid labour and purchase agricultural inputs, locking them into a form of bonded labour. As Leiten and Breman note, "this system of land tenure and labour bondage is the organising principle of agricultural production in Sindh province and of prime importance in explaining the persistent pattern of rural poverty" (2002: 342). Far from being a phenomenon associated with rural 'backwardness' or 'tradition', these relations predominate in the more agriculturally advanced irrigated zones, including parts of southern Punjab, where many landless labourers also remain trapped within debt-driven bonded labour (Martin 2009).

These dynamics stretch across post-colonial Pakistan, remaining firmly embedded in the era of economic liberalisation ongoing since the 1990s. Under dictatorship and then civilian rule, the federal government maintained the elite farmer-orientated Green Revolution technologies, but the ecological windfalls that had accompanied the original strategy have waned (Gadi 2003).

Reemphasising cash crops – including cotton and sugar cane under structural adjustment programmes in the 1990s – had decisively mixed productivity and distributional results, particularly as Green Revolution technologies proved increasingly ineffective on the more marginal lands that had been brought into cultivation (Byerlee and Siddiq 1994; Zaidi 2005). For some, these concerns indicated that the trajectory of Pakistani rural society – despite the agricultural yield improvements realised by the Green Revolution – has reached important social and ecological limits in the form of increasingly stark distributional consequences and forms of environmental degradation (Niazi 2004; Sadeque 2008).

The structural adjustment reforms of the 1990s impacted harshly upon both rural and urban Pakistan. Overall development expenditures retreated from a high of 9.3 per cent of GDP in 1980–81 to two per cent by 2000–01. This was coupled to major reductions in public employment, subsidies to food prices and social expenditures (Zaidi 2005). While the mechanisation of large-scale farming furthered the creation of a rural 'labour surplus' and a renewed process of land concentration accelerated rural outmigration, the urban centres proved strained receptors for migrants. While providing a key livelihood strategy during a decade in which rural poverty sharply increased, rural-to-urban migration brought challenges in terms of access to housing, water and employment. Many of the new peri-urban settlements lack basic services and were located in flood-prone areas (Hasan 2010). Concurrently, migrants from lower classes and castes became heavily dependent on informal-sector work that replicates the patron-client structures of the countryside within which employment is achieved at the cost of affirming new relationships of dependency upon labour contractors (Kabeer et al. 2010; Akhtar 2011). In this context, poverty rose notably over the 1990s and into the early 2000s.

While the above paragraphs highlighted the class and caste dynamics of these relations, other analysts have emphasised the fractures of gender that overlay these dynamics. Notwithstanding important regional differentials and those between rural and urban spheres, the production of gender across rural Pakistan heavily stratifies the roles, responsibilities, resources and power accorded to men and women. The latter often face restricted access to private resources (land, credit), public goods (health, education) and encounter patriarchal barriers to influence upon decision-making within and outside the household (Shahbaz 2008; Kabeer, Mumtaz and Sayeed 2010). These gendered relations manifest themselves in pervasive gender inequalities in terms of health, nutrition, education and political voice, which at intersections of gender, class and caste act as vectors of insecurity that create profound vulnerability for women in Pakistan. This is particularly notable in the way that women and girls often face the brunt of food insecurity, as gendered relations unequally shape the distribution of already scarce household resources (Kugelman 2010). Fortunately, Pakistan's national planning commission had a suitable response to the ills of malnutrition: in 2002 it reduced the reference threshold in determining the national poverty line from a minimum calorie intake of 2,550 per adult per day to 2,350 per adult per day as required to maintain physical functioning and daily activities. If poverty and malnutrition indicators stubbornly refused to go down, it seemed then that the metrics used to measure them could be lowered so as to create the same effect.

The 2010–2011 floods as a window on rural vulnerability

The above paragraphs sketched the historical lineage of agrarian social relations within contemporary Pakistan, emphasising how the experience of vulnerability is produced through the relational dynamics between different social groups as expressed within the socio-ecological setting of a produced rural environment. Across the Indus watershed, ownership of land stands at the nexus of a series of power relations involving control over water, credit and the labour of others. These social relations are central to the internal constitution of vulnerability as a socio-ecological relationship and greatly shape the ability of different social groups to adapt to hydroclimatic variations. Conceptualising vulnerability in this manner indicates how what is commonly termed 'climate change adaptation' is fundamentally a relational process. It is predicated upon the power of individuals and communities to change or maintain specific livelihoods in the face of changing socio-ecological conditions, including their forms of interaction with other social groups and classes. The struggle to exercise control over key productive assets is pivotal for households and social groups to minimise adverse impacts and to avoid having other groups displace insecurities onto them.

The experience of the 2010 and 2011 floods emphasised precisely how the hydroclimatic shifts linked to anthropogenic climate change express themselves through this historically constructed socio-ecology of agrarian Pakistan. The notion of 'internal defenselessness' to an external climate, despite underpinning the national adaptation strategy, is grotesquely misplaced in a context of agrarian relations predicated precisely on disempowerment to uphold a brutally unequal rural order. Two examples help to accentuate these relational dynamics. First, the disruption of agricultural livelihoods in the post-flood aftermath made vulnerable groups expressly dependent upon sources of credit to survive. Aid workers found that, after satisfying immediate food needs, the second most prominent manner of spending government instigated post-disaster cash transfers was to pay the interest on loans to local moneylenders, merchants and landlords (CARE International 2012). Recipients of the cash transfers knew that such emergency assistance would be short-lived. They recognised that future sources of credit – for both immediate consumption and agricultural production – would come from habitual informal credit sources to whom they were already indebted. As such, poor households understood that it was expressly important to reassert creditworthiness within the informal credit market despite this being a key source of their exploitation and dependency. The frequency of new debts increased 10 times overall as a direct result of the floods, with the overall value of debt increasing by 78 per cent (CARE International 2012: 5). This escalation of indebtedness provided landowners and merchants with an expanded source of monetary income and social power to help overcome their own disruptions to production and labour relations. For landless labourers, tenants and smallholders, however, increasing their indebtedness reaffirmed existing social relations of dependency and exploitation. To them, however, this was a necessary manner of dealing with immediate vulnerabilities structurally ingrained by the inequities of ownership and power within rural Pakistan.

Second, landless tenants and labourers are habitually dependent upon landlords to allow them to construct houses upon their lands. This leaves them expressly vulnerable to floods because, as Mustafa (2002) noted, such houses are habitually built on exposed land. Moreover, it also reinforces dependency upon the landlords who have a right to eject them. Together, this double vulnerability is self-reinforcing because uncertain land tenure undermines the willingness of tenants to invest in the physical infrastructure of their houses, thereby reinforcing vulnerability to the elements. In the post-flood aftermath, however, high local and international food prices created incentives for landowners to increase cropped land. As a consequence, for some landlords the floods represented a welcome process of creative destruction within the lived environment. Now that settlements had been washed away, land that previously had been occupied by tenants could be more easily reclaimed for agricultural purposes (Budhani and Gazdar 2011). Indeed, despite the floods creating massive social disruption born primarily by the landless and small cultivators, agricultural earnings during the period actually expanded as large landholders exported crops to take advantage of high international prices for agricultural commodities (Iqbal 2011). This may in part be related to widely reported instances of landlords using political leverage to direct flood waters away from their agricultural lands and onto the lands occupied by settlements or smallholder groups (Mustafa and Wrathall 2011; Breman 2013).

The destruction of the lived environment through floods also opened up some potential openings for marginalised social groups to insert themselves into different livelihood trajectories. In destroying the social and physical infrastructures that kept them tied into vastly unequal agrarian relations, the floods provided a moment of rupture that temporarily loosened the bonds that fixed marginal bodies within the rural landscape. Many *hari* sharecroppers from Sindh, for example, sought to use the destruction of the floods to escape the debt bondage that tied them to landlords. In the period of confusion and reorganisation that accompanied the floods, individuals and families began to uproot to Karachi and other urban centres. Indeed, the prospect of migration chipping away at the subordinate sharecropping labour force was not lost on the Sindhi landlord class, who used their considerable political leverage to have the camps for those displaced by floods shut down to prevent them serving as a launch pad for urban migration (Breman 2013). For those that did escape, however, this flight from oppression entailed an uncertain liberation into a desperately poor and insecure future at the margins of urban existence. As Jan Breman describes, "[h]alfway between the countryside and the city they get stuck in a kind of nowhere landscape" (2013: 39). Their futures, therefore, become interwoven with the contradictions of surplus bodies situated at the margins of urban Pakistan, a prospect that raises a daunting new set of challenges and constraints for such marginalised classes (Hasan 2002, 2010).

Conclusion

This chapter has drawn on the tools set out in the book to begin the process of historicising climate change adaptation within the longer-term socio-ecological transformations of rural Pakistan. As the post-2010 flood assessments reveal, recovery from natural hazards is filtered through relationships of credit/debt and control over land through which dominant social classes are able to consolidate their position within the rural hierarchy. To do so is to both monopolise gains that the socio-ecology of an intensely canalised agrarian environment provides, while displacing the inherent risks onto others. Despite the importance of these power relations for understanding the possibilities of and limitations to climate change adaptation strategies, they are entirely marginalised in the Pakistani government's technical and managerial approach to adaptation. Within the latter, the discourse of climate change adaptation in Pakistan has been constructed in singularly Malthussian terms as a struggle against natural limits and threats. The latter facilitates the conceptualisation of 'adaptation' as a technical exercise that stands external to the ongoing processes of agrarian transformation that hinge on the exercise of power bound up with the control of land, water, labour and debt.

To reaffirm, there is undoubtedly a need for managerial reforms and technical solutions to mediate the adverse impacts of hydroclimatic shifts and related events such as floods and water shortages. Equally, these must be complemented by community-based mobilisation to address specific vulnerabilities on a micro-level by increasing the 'adaptive capacity' of marginalised groups (see Ensor and Berger 2009). Yet the analysis cannot begin or stop there because the possibilities and limits to both macro and micro adaptation strategies are decisively shaped by the context of historically structured power relations that produce the agrarian environment. The possibilities for social justice, therefore, are inseparable from a broader transformation of the region's agrarian structure and its socio-ecological underpinnings. In this respect, vast inequities over the control of key productive assets remains the elephant in the room of climate change adaptation discourse and policy in Pakistan. This is not surprising given that – as Chapters Three and Four argued – they are striking absent from the discourse of climate change adaptation more generally.

Historicising climate change raises the prospect that a serious conversation about what is termed 'climate change adaptation' – in Pakistan and other agrarian environments – will necessarily raise the issues of land reform and the equitable management of water resources, notwithstanding the daunting political challenges they raise given the concentrated control over the levers of state power assumed by elite classes. In this respect, the Pakistani rural poor do not suffer from any difficulties in identifying the elephant in the room. Surveys conducted as part of the 2003 Pakistan National Human Development Report clearly indicated that marginalised groups overwhelmingly considered the lack of access to land – either through ownership or secure usufruct rights – was central to their condition of poverty and disempowerment (Hussain 2003). Although it is glaringly absent in current strategies, land reform might be considered a 'climate change adaptation' policy par excellence in the Pakistani context. Michael Lipton argues in his recent broad study of land reform in developing countries that, notwithstanding political opposition, land reform in Pakistan remains a potential source of agrarian revival that would substantially reduce rural poverty and inequality with likely net gains to production (Lipton 2009: 292; see also, Zaidi 2005). There are, of course, different ways to envisage the possibilities and limits to land reform (Bernstein 2002), yet putting control over land at the forefront of climate change adaption is a silence that needs breaking. It would be one way to move the policy discourse beyond the platitudes to 'pro-poor growth' that have steadfastly proved unable to change the vast inequalities of contemporary rural Pakistan (see Lieten and Breman 2002).

The current federal government in Pakistan, however, explicitly avoids any mention of land issues in the National Climate Change Policy. At the time of writing, its conception of land politics appears quite different, having actively launched a controversial package to lease up to seven million acres of state land to foreign investors. The latter is a process that appears to be at odds with its constitutional pledge to enact equitable land reforms (Settle 2012). Despite this, in other institutional locations there exist initiatives that indicate the continued salience of the land issue in the context of unsettled rural politics. In 2008, for example, under pressure from various social movements and NGOs, the Sindh provincial government launched a programme to distribute government lands to female heads of landless households. The stated rationale for this programme is notable: land ownership was projected as the primary means of addressing vulnerability, and targeting women simultaneously addressed a second vector of disempowerment. Although the programme was relatively small-scale and encountered significant design and implementation troubles during its initial implementation (see, PDI 2009: for a review), the renewed linking of land to questions of poverty and empowerment creates an important precedent in the way that it exceeds the discursive constraints of climate change adaptation.

All this indicates how climatic change is becoming a new institutional arena for political contestation in which different social agents attempt to ascribe particular visions of change, security, justice and efficiency to the idea of adaptation. While this certainly offers certain possibilities for progressive outcomes, it nonetheless occurs within a discursive frame that constructs climate change adaptation in a manner that pushes such issues to the margins. To draw out these trends more fully, it is useful to move from the macro-historical level to a more finely graded examination of the dynamics of hydrological cycles, agricultural production and power at a regional level. To do so, the following chapter draws out processes of agrarian transformation, common property resources and the uneven distribution of risk from climatic variability in context of the semi-arid regions of the Deccan Plateau in southern India.

Notes

1 The Pakistani government had first projected a climate change adaptation strategy in 2002, yet this was not well integrated into substantive policy practice (Khan 2011). Notably, James Wescoat (1991) had raised the issue of mapping climate change impacts across the Indus watershed in a prescient 1991 piece.

2 For space purposes, agrarian change in the Mughal period preceding British colonialism is left aside here; see Nazir (2000) and Rahman (2013).

3 Displacement is another key factor. Construction of the Tarbela dam – the centrepiece of the new Indus engineering projects – displaced more than 100,000 people (Gadi 2003).

References

Akhtar, A. 2011. "Patronage and class in urban Pakistan." *Critical Asian Studies* 43(2): 159–84.

Akram-Lodhi, H. 2001a. "'Like an act of god': Land, water and social power in northern Pakistan." *Contemporary South Asia* 10(3): 319–42.

Akram-Lodhi, H. 2001b. "'We earn only for you': Peasants and 'real' markets in Northern Pakistan." *Capital & Class* 25(2): 79–110.

Ali, I. 2002. "The historical lineages of poverty and exclusion in Pakistan." *South Asia: Journal of South Asian Studies* 25(2): 33–60.

Archer, D., N. Forsythe, H. Fowler, and S. Shah. 2010. "Sustainability of water resources management in the Indus Basin under changing climatic and socio economic conditions." *Hydrology and Earth Systems Sciences* 14(8): 1669–80.

Bankoff, G. 2001. "Rendering the world unsafe: 'Vulnerability' as western discourse." *Disasters* 25(1): 19–35.

Bernstein, H. 2002. "Land reform: Taking a long(er) view." *Journal of Agrarian Change* 2(4): 433–63.

Bhutta, M., and E. van der Velde. 1992. "Equity of water distribution along secondary canals in Punjab, Pakistan." *Irrigation and Drainage Systems* 6(2): 161–77.

Breman, J. 2013. "Land flight in Sindh." *Economic and Political Weekly* XLVII(9): 35–40.

Briscoe, J., U. Qamar, and M. Contijoch. 2005. *Pakistan's Water Economy: Running Dry*. Washington, DC: World Bank Group.

Budhani, A., and H. Gazdar. 2011. "Land Rights and the Indus Flood, 2010–2011: Rapid Assessment and Policy Review." Edited by Oxfam Research Report. Karachi: Oxfam.

Byerlee, D., and A. Siddiq. 1994. "Has the green revolution been sustained? The quantitative impact of the seed-fertiliser revolution in Pakistan revisited." *World Development* 22(9): 1345–67.

CARE International. 2012. *Rapid Assessment: Indebtedness in the Flood-Affected Regions of Pakistan*. Karachi: CARE International in Pakistan.

Cheesman, D. 1981. "Power in rural Sind in the late nineteenth century." *Asian Affairs* 12(1): 57–67.

Cheesman, D. 1982. "'The omnipresent Bania:' Rural moneylenders in nineteenth-century Sind." *Modern Asian Studies* 16(3): 445–62.

Cruz, R., H. Harasawa, M. Lal, S. Wu, Y. Anokhin, B. Punsalmaa, Y. Honda, M. Jafari, C. Li, N. Huu Ninh, C. Cambridge, and University Press. 2007. "Asia – Climate Change 2007: Impacts, Adaptation and Vulnerability." N.p. in *Contribution of Working Group II to the Fourth Assessment Report of the Intergovernmental Panel on Climate Change*, edited by M. L. Parry, O. F. Canziani, J. P. Palutikof, P. J. van der Linden, and C. E. Hanson. Cambridge: Cambridge University Press.

Davis, M. 2002. *Late Victorian Holocausts*. London: Verso.

Ensor, J., and R. Berger. 2009. *Understanding Climate Change Adaptation: Lessons form Community-Based Approaches*. Rugby, UK: Practical Action Publishing.

Gadi, M. 2003. "Counter narrative of irrigation management: The case of Pakistan." *Water Nepal* 9(1/2): 269–76.

Gazdar, H. 2011. "The Fourth Round, and Why They Fight On: The History of Land Reform in Pakistan." Pp. 8–65 in *Leveling the Playing Field: A Survey of Pakistan's Land Reforms*, edited by Ali Cheema. Kathmandu: Panos South Asia.

Gilmartin, D. 1988. *Empire and Islam: Punjab and the Making of Pakistan*. London: Tauris.

Government of Pakistan. 2011. "National Climate Change Policy." Edited by Ministry of Environment. Lahore: Government of Pakistan.

Government of Pakistan. 2012. "Pakistan's Pathway to a Sustainable & Resilient Future." Edited by Ministry of Climate Change. Islamabad: Government of Pakistan.

Gronewold, N. 2010, August 18. "Is the flooding in Pakistan a climate change disaster." In *Scientific American*. Available at: http://www.scientificamerican.com/article/is-the-flooding-in-pakist/

Haines, T. D. 2011. "Building the Empire, Building the Nation: Water, Land and the Politics of River-Development in Sind, 1898–1969." PhD dissertation, Royal Holloway College, London, University of London.

Hamid, N. 1982. "Dispossession and differentiation of the peasantry in the Punjab during colonial rule." *Journal of Peasant Studies* 10(1): 52–72.

Haq, R. 2007. "Land inequality by mode of irrigation in Pakistan, 1990–2000." *Pakistan Development Review* 46(4): 1011–22.

Hasan, A. 2002. "The changing nature of the informal sector in Karachi as a result of global restructuring and liberalization." *Environment and Urbanization* 14(1): 69–80.

Hasan, A. 2010. "Migration, small towns and social transformations in Pakistan." *Environment and Urbanization* 22(1): 33–50.

Hussain, A. 1988. *Strategic Issues in Pakistan's Economic Policy*. Karachi: Progressive Publishers.

Hussain, A. 2003. *Pakistan: National Human Development Report 2003*. Karachi: UNDP/ Oxford University Press.

ICUN (International Union for Conservation of Nature), Ministry of Environment, and DIFD. 2011. "Climate Change Vulnerabilities in Agriculture in Pakistan." Karachi: IUCN.

IPCC. 2001. *Climate Change 2001: Impacts, Adaptation and Vulnerability. Summary for Policy Makers*. Geneva: World Meteorological Organisation.

Iqbal, S. 2011, May 7. "Farm sector earns more despite floods." *DAWN*. Available at: http://www.dawn.com/news/626564/farm-sector-earns-more-despite-floods

Joshi, P. C. 1974. "Land reform and agrarian change in India and Pakistan since 1947." *Journal of Peasant Studies* 1(2): 164–85.

Kabeer, N., K. Mumtaz, and A. Sayeed. 2010. "Beyond risk management: Vulnerability, social protection and citizenship in Pakistan." *Journal of International Development* 22: 1–19.

Khan, M. 2007. "The Political Ecology of Irrigation in Upper Sindh: People, Water and Land Degradation." PhD dissertation, Department of Anthropology, Binghamton, NY, Binghamton University, SUNY.

Khan, M.A.A. 2011. "National Economic, Environment and Development Study (NEEDS)." Edited by Ministry of Environment. Islamabad: Government of Pakistan.

Kugelman, M. 2010. "Pakistan's Food Insecurity: Roots, Ramifications, and Responses." Pp. 5–30 in *Hunger Pains: Pakistan's Food Insecurity*, edited by Michael Kugelman and Robert Hathaway. Washington, DC: Woodrow Wilson International Centre.

Kugelman, M., and R. Hathaway (Eds.). 2009. *Running on Empty: Pakistan's Water Crisis*. Washington, DC: Woodrow Wilson International Center.

Lieten, K., and J. Breman. 2002. "A pro-poor development project in rural Pakistan: An academic analysis and a non-intervention." *Journal of Agrarian Change* 2(3): 331–55.

Lipton, M. 2009. *Land Reform in Developing Countries: Property Rights and Property Wrongs*. New York: Routledge.

Martin, N. 2009. "The political economy of bonded labour in the Pakistani Punjab." *Contributions to Indian Sociology* 43(1): 35–59.

Ministry of Finance. 2012. "Pakistan Economic Survey 2011–12." Edited by Ministry of Finance. Islamabad: Government of Pakistan.

Mishra, S. C. 1982. "Commercialisation, peasant differentiation and merchant capital in late nineteenth-century Bombay and Punjab." *Journal of Peasant Studies* 10(1): 3–51.

Mitchell, T. 2002. *Rule of Experts: Egypt, Techno-Politics, Modernity*. Berkeley: University of California Press.

Mustafa, D. 2001. "Colonial law, contemporary water issues in Pakistan." *Political Geography* 20(4): 817–37.

Mustafa, D. 2002. "To each according to his power? Participation, access, and vulnerability in irrigation and flood management in Pakistan." *Environment and Planning D: Society and Space* 20(3): 737–52.

Mustafa, D. 2005. "The production of an urban hazardscape in Pakistan: Modernity, vulnerability, and the range of choice." *Annals of the Association of American Geographers* 95(3): 566–86.

Mustafa, D. 2010. "Hydropolitics in Pakistan's Indus Basin." United States Institute of Peace Special Report #261. Washington, DC: United States Institute of Peace.

Mustafa, D., and D. Wrathall. 2011. "Indus basin floods of 2010: Souring of a Faustian bargain?" *Water Alternatives* 4(1): 72–85.

Nazir, P. 2000. "Origins of debt, mortgage and alienation of land in early modern Punjab." *Journal of Peasant Studies* 27(3): 55–91.

Niazi, T. 2004. "Rural poverty and the green revolution: The lessons from Pakistan." *Journal of Peasant Studies* 31(2): 242–60.

PDI (Participatory Development Initiatives). 2009. *Sindh Government's Land Distribution Programme: Issues and Challenges*. Karachi: PDI/Oxfam.

Qureshi, A., P. McCornick, A. Sarwar, and B. Sharma. 2010. "Challenges and prospects of sustainable groundwater management in the Indus Basin, Pakistan." *Water Resources Management* 24(8): 1551–69.

Rahman, T. 2013. *The Class Structure of Pakistan*. Karachi: Oxford University Press.

Sadeque, N. 2008. "Gambling on Pakistan's agricultural future." *Development and Change* 51(4): 555–60.

Settle, A. 2012. "Agricultural Land Acquisition by Foreign Investors in Pakistan: Government Policy and Community Responses." Land Deal Politics Initiative Working Paper #7. The Hague: International Institute of Social Studies.

Shahbaz, B. 2008. "Risk, Vulnerability and Sustainable Livelihoods: Insights from Northwest Pakistan." Edited by Sustainable Development Policy Institute. Islamabad: Sustainable Development Policy Institute.

Stern, N. 2007. *The Economics of Climate Change: The Stern Review*. Cambridge: Cambridge University Press.

Washbrook, D. 1993. "Economic depression and the making of 'traditional' society in colonial India, 1820–1855." *Transactions of the Royal Historical Society* 6(3): 237–63.

Wescoat, J. 1991. "Managing the Indus River basin in light of climate change: Four conceptual approaches." *Global Environmental Change* 1(5): 381–96.

Wescoat, J., S. Halvorson, and D. Mustafa. 2000. "Water management in the Indus Basin of Pakistan: A half-century perspective." *Water Resources Development* 16(3): 391–406.

Zaidi, S. A. 2005. *Issues in Pakistan's Economy (2nd edition)*. Karachi: Oxford University Press.

7 India – water, debt and distress in the Deccan plateau

In the previous chapter, I examined the historical context of climatic change and socio-ecological transformations across the Indus watershed in Pakistan at a broad level. The present chapter shifts to a more focused regional level of analysis to examine agrarian environments in the Deccan plateau in south-central India. The Deccan constitutes a semi-arid region that stretches across the heartlands of peninsular India in which around 60 per cent of the population derive their livelihoods from activities related to agriculture. With the region located in the rain shadows of the Western and Eastern Ghats, rainfall typically is scarce for most of the year and the summer monsoons are often fickle. Given that the livelihood strategies of a majority of households remain directly or indirectly tied to agricultural activities, multiple concerns have been raised about the potential adverse impacts of climate change. In a region already prone to regular and sometimes severe droughts, there are explicit fears that increasingly erratic rainfall patterns will have notable adverse effects upon the agrarian economy in general and smallholder cultivators in particular.

Such fears are well grounded. Studies of the region's hydro-climatic record over the past four decades indicate that climate change effects already are being experienced in the form of rising overall temperatures, more variable and extreme rainfall patterns and an increasing frequency and severity of drought (Lambrou and Nelson 2010). Discerning a secular trend from decadal variability is difficult, yet the results tend to conform to projections of climatic change made by several sources. The World Bank predicts substantially higher temperatures for semi-arid Andhra Pradesh, a state on the eastern side of the Deccan plateau, with a slight increase in quantity of rain yet more erratic patterns of rainfall (World Bank 2008: 5–6, 42–44; see also, Pande and Akermann 2010). This volatility is expected to translate into greater frequency of drought years, mixed with years of above average rainfall in which precipitation is concentrated, therein much of the extra water may be lost to runoff.[1] As a consequence, the World Bank predicts a dismal impact upon agriculture for the state:

> With high prevailing baseline temperatures these changes generate deteriorating agroclimatic conditions, with declining yields for the major crops (rice, groundnut, and jowar) . . . In the harsher climate change scenarios,

farm incomes could decline substantially (by over 20%), suggesting that agriculture as currently practiced may not be capable of sustaining large populations on small rain-fed farms.

(World Bank 2008: 42)

The paradox of climate change adaptation in the semi-arid Deccan, however, is that agriculture as currently practiced on small farms is already in the midst of an ongoing crisis. Over the last two decades, smallholders across semi-arid regions of agrarian India have faced tightening constraints on profitability, escalating levels of indebtedness and deepening constraints upon their means of social reproduction. The latter has been most tragically instantiated in the wave of farmer suicides that have plagued semi-arid areas. At a national level, around 250,000 farmers faced with crippling debts have taken their lives over the past 15 years, the largest wave of recorded suicides in human history (Sainath 2010a). Roughly half of those suicides have been concentrated in the central Deccan region (see Table 7.1).

These suicides are a tragic instantiation of a dramatic level of human insecurity across the region. In part, they are related to biophysical processes wherein crop failure occurs due to absent or erratic rains, which creates compounding pressures upon households. Yet, as numerous observers have noted, droughts have been a longstanding feature of this agrarian environment whereas mass farmer suicides are a much more recent phenomenon. They do not simply reveal the unarguable stresses posed upon households by drought. Rather, they open a window onto a

Table 7.1 Number of suicides per year in central Deccan region

Year	*Andhra Pradesh*	*Maharashtra*	*Karnataka*	*All India*	*Deccan 3 as % of India total*
2003	1800	3836	2678	17164	48
2004	2666	4147	1963	18241	48
2005	2490	3926	1883	17131	48
2006	2607	4453	1720	17060	51
2007	1797	4238	2135	16632	49
2008	2105	3802	1737	16196	47
2009	2414	2872	2282	17368	44
2010	2525	3141	2585	15964	52
2011	2206	3337	2100	14027	54
Total	**20610**	**33752**	**19083**	**149783**	**49**
Yearly Average	***2290***	***3750***	***2120***	***16643***	

Data drawn from (Sainath 2012).

new dynamic of insecurity that has arisen in the context of severe cost pressures stemming from liberalised agriculture, escalating levels of household debt and the transformation of the socio-ecological infrastructures through which farmers seek access to irrigation.

In what follows, I seek to trace out this relationship between water, debt and agrarian distress with a focus on semi-arid Andhra Pradesh. While agriculture in such regions may well remain in part a 'gamble with the monsoons', it is nonetheless a gamble that is expressed through engrained power relationships through which actors attempt to exercise control over land, hydrology and credit. I emphasise how, owing to limited access to alternative forms of irrigation, farmers in semi-arid regions have increasingly taken on debts to drill borewells that access water from underground aquifers. At first, this debt-driven form of irrigation appeared as an important strategy for smallholders to overcome the tightening margins of production in a period of austere agrarian liberalisation. Harnessing these new flows of water helped remove constraints to the use of Green Revolution technologies, including HYV seeds that were expected to improve yields and alleviate pressures upon rural incomes and livelihoods. By reworking the region's hydrology, the very climate of agricultural production changed. Notwithstanding such expectations, however, in many semi-arid areas the competitive drilling of wells has created a race to the bottom of the water table. In these cases, the initial windfalls provided by groundwater irrigation receded under conditions of acute resource exhaustion. The latter has been manifested in aquifer depletion, water contamination and an agrarian environment littered with failed wells, withered crops and intractable debts. As such, the ubiquity of debt and the scarcity of water are parallel and deeply intertwined socio-ecological dynamics. The possibilities and limits to 'climate change adaptation', therefore, can only be understood within this socio-ecological context of an adapted agrarian climate.

The political ecology of irrigation and debt

A persistent struggle for control over water has been central to India's history since settled agriculture. In part, the diverse attempts to shape or enclose its fugitive character appear as an enduring element of human struggles to produce lived environments. From its unreliable birth in the form of rain or snow, to its movements across the land and underground through rivers, streams and aquifers, water has a fickle presence that greatly challenges agricultural endeavours. Using techniques ranging from hand-dug ditches, to local wells to audacious macro-projects of hydrological engineering, humans have repeatedly attempted to incorporate the uneven temporal and spatial presence of water for their own purposes. The resulting socio-ecological infrastructures, however, emerge not only as technical preconditions to agriculture but also as a core nexus in the power relations that animate the agrarian environment. As David Ludden (1999: 37) has argued, irrigated agriculture in South Asia incorporates not simply the physical distribution of surface and sub-surface water but also, inseparably, the institutional configurations that source water onto farms and the movement of ideas about technology, power and justice that make those institutions work. These social institutions, physical structures and cultural practices through which the collection and distribution of water is accomplished – a 'waterscape' in the terms of Amita Baviskar (2007) – have historically been interwoven with prevailing hierarchies of caste, class and gender. Social power and water, therefore, are intimately intertwined: power is deployed to secure access to water and access to water reinforces structures of power (cf. Swyngedouw 2004).

On this basis, the socio-ecology of water is intimately wedded to how agricultural surpluses are produced and expropriated and how the risks of production are unevenly distributed between social groups within and beyond the immediate locality. Complex socio-ecological relations of this type are not static nor are they driven by strictly localised dynamics. On the contrary, these fluid relations are continuously refashioned within broader historical trends, including changing state formations and the shifting integration of local production and consumption into regional and global divisions of labour. Successive forms of integration into wider political economic dynamics – from the Mughal empire, through British colonialism, to national developmentalism and currently liberalised agriculture – have driven repeated processes of transformation and contestation (Gadgil and Guha 1993, 1994; Arnold and Guha 1995; Mosse 2003; Hill 2008).

In the context of contemporary semi-arid regions of southern India, such 'waterscapes' have become increasingly interlocked with what we could correspondingly term 'debtscapes'. Here I emphasise the social relations of credit/debt that permeate the agrarian environment and structure key social relations surrounding agrarian life. Access to credit appears as a pivotal means for farmers to overcome barriers to production and profitability. By facilitating investments in land, irrigation, inputs and labour, credit can shift the socio-ecological parameters of agricultural production, potentially allowing for an expansion of cropped area or an increase in the capital intensity of production. On this basis, state programmes have often sought to expand formal credit into rural areas as a lever of agrarian modernisation (Shah, Rao and Shankar 2007).

Concurrent to its productive role, credit from either formal or informal sources also is central to the livelihood strategies of rural households. The latter repeatedly navigate the varied temporal horizons of agrarian life, in which expenses for various productive and consumptive tasks do not necessarily align with the seasonal flow of incomes derived from harvests, off-farm employment, remittances from migration and other livelihood strategies. Unexpected expenses from illness to crop failure exacerbate such tensions, and credit may appear the only means to avoid distress sales of key assets in such circumstances. To address these risks, households commonly invest considerable material and cultural resources into maintaining multiple sources of credit from family, friends and community members, despite this often requiring the reaffirmation of existing social hierarchies (Guérin et al. 2011). On this basis, the extension of formal credit into the realm of household reproduction is repeatedly evoked as an important tool of rural development, as recently evidenced in the paeans to the transformative potential of microcredit and the accompanying discourse of 'financial inclusion' (see Taylor 2012).

Notwithstanding such modernising aspirations, the role of credit in transforming rural social relations has been highly uneven. Taking on credit, whether to satisfy immediate consumption needs or to invest in well drilling or other productive activities, provides a temporal shift of risk into the future to a point when the loan must be repaid. This displacement frequently comes at the cost of re-establishing relations of exploitation and dependency with the provider of credit. Relationships of credit, therefore, are simultaneously and intractably relations of debt, involving an immediate and often enduring power relationship between creditor and debtor. As a significant body of work on the political economy of agrarian transformation in India has documented, relationships of debt are strongly interwoven into the class and caste hierarchies of rural life and serve as key means of expropriating surpluses from producers by landlord, merchant and moneylender classes (Olsen 1996: provides a classic study in Andhra Pradesh). In particular, the need for smallholders to repeatedly borrow money to purchase agricultural inputs at the beginning of the season often locks them into lopsided debt relations with landed and merchant capital in which the power of credit is used to bind producers into relationships that manifestly favour the latter. These kinds of relationships have been integral to the workings of agrarian capitalism, as discussed in Chapter Four, and have long historical lineages within the Deccan region (see, in particular, Banaji 1977).

As such, while credit appears as a necessary tool for household livelihoods that enables households to push back against constraints upon social reproduction, it simultaneously amplifies the scale of risks faced by borrowers and can entrench localised hierarchies predicated upon exploitation and dependency. These contradictory dynamics are particularly evident in the context of current agrarian transformations within semi-arid Andhra Pradesh. Here credit became the preponderant means for financing the drilling of borewells through which smallholders sought access to groundwater for irrigation. The resulting 'groundwater revolution' not only underlay a quantitative extension of irrigated area but also provided hydrological foundations for social and technological change across semi-arid regions. A new socio-ecology of the agrarian environment was being produced as, purportedly, the foundation for a resurgence of rural prosperity. As the following sections indicate, however, the aspirations for agrarian modernisation that accompanied these shifts fractured in the context of compounding socio-ecological contradictions. For smallholders across these regions debt remains an unwavering companion of insecurity. This greatly complicates how we conceptualise climatic change within the region.

The anatomy of the 'groundwater revolution'

Jawaharlal Nehru, India's first prime minister, famously labelled large dams the 'new temples' of modern India owing to their role in mastering the flows of water essential for agricultural irrigation. In this era of national developmentalism, control over water was posited as synonymous with India's control over its own destiny. Macro-irrigation projects predicated upon barrages, river diversion and canalisation signified a commitment to scientific rationalism, the taming of climate and the promise of a productivity revolution in agriculture. Over the past three decades, however, the grand temples of large-scale canalisation have been usurped by the tiny shrines of millions of individual groundwater pumps. Aided by state provisions for subsidised electricity alongside new drilling and pump technologies, the Indian countryside has witnessed the proliferation of private, pump-driven borewells that source water from underground aquifers. Under this 'groundwater revolution', India has become the biggest user of groundwater globally. Some 20 million wells currently source water for almost 60 per cent of India's net irrigated area (Shah 2007).

As with canalisation, the attractiveness of groundwater irrigation lay precisely in its potential to sever agricultural production from the fickleness of the monsoon rains. The possibility of 'water on demand' provided hydrological foundations for a greater intensity of production through year-round cultivation, the diversification of crops and increased technological change across both production and processing. In particular, by rendering flows of water through the agricultural landscape more malleable to human use, groundwater abstraction appeared to unlock important socio-ecological barriers to the generalisation of Green Revolution technologies. The latter, predicated upon HYV seeds used in conjunction with synthetic fertilisers and pesticides, are strongly dependent upon regular and sustained infusions of water for germination and growth. Prior to the widespread drilling of borewells, the constraint of water access had made the usage of such technologies extremely tenuous outside of the canal-irrigated zones.

This convergence between the extension of groundwater irrigation and the generalisation of commercially orientated agriculture – founded upon the intensive use of productivity-enhancing technology – has repeatedly been heralded as a significant driver of rural development and poverty alleviation. While canal irrigation was geographically constrained and strongly influenced by national and regional politics, any farmer who could afford to hire a drilling rig and install and maintain a pump could in principle tap into underground aquifers for in-situ water-on-demand. This led to claims of a 'democratisation of irrigation', with smallholder agriculturalists able to procure water despite their frequent marginalisation within canalised zones. Not only was groundwater envisaged to augment household incomes by increasing agricultural productivity and a switch to higher-value crops, it was also deemed central to a broader revitalisation of the rural economy by buttressing the extent of associated activities such as processing, transportation and marketing (Mukherji and Shah 2005). For agro-economist Tushaar Shah, groundwater abstraction constituted a poverty alleviation tool par excellence:

> It can be safely said that the groundwater boom has been amongst the best things that have happened for South Asia's rural poor in the past few decades, and the size and dispersion of the livelihood benefits of this boom can arguably outcompete some of the best-known poverty alleviation programmes in the region.
>
> (Shah, 2007: 32)

Groundwater therein promised a form of agrarian modernisation and poverty alleviation without tackling longstanding issues of unequal access to land and other productive assets. Under the presupposition that borewells represented a depoliticised technology, state bodies at both national and regional levels heavily promoted groundwater extraction. Whereas projects aimed at increasing the scope of canal irrigation have slowed or stagnated across most of rural Indian, successive governments at regional and national levels have encouraged private well drilling through the extension of credit and also by subsidising the electricity used by pumpsets. In Andhra Pradesh, for example, a central pillar of the current Congress government's agrarian populism has been the promise of providing free electricity to farmers for seven hours a day specifically for pumping purposes. This policy was justified as an equity measure undertaken to expand access to groundwater within a smallholder sector chastened by agricultural liberalisation policies since the 1990s. Moreover, given that canal irrigation has largely bypassed the semi-arid interior areas such as Telangana and Rayalaseema, the policy was presented as a way to redress historical grievances over irrigation schemes (Nag 2011). Subsidised groundwater abstraction, therein, was posited as both a means of overcoming agrarian distress and alleviating longstanding uneven development across the state.

Notwithstanding the great expectations placed upon groundwater as a tool of rural development, its impact upon poverty and vulnerability has been decisively uneven. In Andhra Pradesh, the earlier optimism attached to the groundwater revolution has been strongly tempered by concerns about an overexploitation of aquifers that threatens not just agricultural production but also the stability of drinking water sources in rural areas. There are multiple interwoven socio-ecological factors behind this apprehension. In the case of the Deccan plateau, upon which interior Andhra Pradesh is situated, the specific geo-hydrology of the region greatly constrains the technical capacity to abstract groundwater on a large scale. In contrast to the sizeable alluvial aquifers that underlie the plains of northern India, the geological composition of the Deccan region's hard-rock formations gives rise to small groundwater aquifers that collect unevenly and shallowly within small fissures and joins. This means that the presence of groundwater varies greatly according to the localised weathering of the rock formation, and drilling for wells involves a gamble that even geological surveying – which is beyond the price range of most smallholder farmers – can only moderate.[2] Under these conditions, aquifers in similar settings elsewhere tend to be used only for drinking water supply, meeting pastoral demands or limited industrial use (World Bank 2010: 20). Across the semi-arid regions of southern India, however, they have become widely exploited for agricultural irrigation.

Compounding the uneven presence of water, even when an aquifer is successfully tapped, the inherent shallowness of underground reservoirs within these rock formations and their dependence on rainwater for recharge make aquifers susceptible to rapid depletion during the course of a season. This is particularly prevalent when over-exploitation – i.e. extraction at a rate beyond recharge – is combined with periods of below-average rainfall, with the latter two processes tending to be mutually exacerbating. Drought increases the pressure upon cultivators to expand groundwater abstraction to compensate for the absence of rain. Simultaneously, drought deprives aquifers of the rainwater necessary for recharge, thereby leading to rapid overexploitation, well failure and the withering of crops. In this manner, climatic variability takes shape in and through both geophysical and social character of the lived environment (see Chapter One).

Two immediate consequences of the dramatic expansion of groundwater use can be noted. First, the shift to private borewell drilling has profoundly disrupted existing forms of irrigation based on the collective use and management of tank systems. These forms of 'minor irrigation', historically preponderant in semi-arid regions, involve the capture of rain runoff and seasonal streams within village-managed tanks of various sizes. Water subsequently can be released for irrigation purposes as determined by social institutions that manage such collective resources at a village level. While the social organisation of tank irrigation often incorporated some aspects of reciprocal resource use across the village setting, these dynamics were commonly shaped by the hierarchical relations predicated upon caste, class and gender that permeated rural society (Janakarajan 1993; Mosse 2003).

These collective forms of tank irrigation, however, have been in decline since the 1950s, in part because the deepening commercialisation of agriculture placed strains upon the social institutions and physical infrastructures underpinning such water-management institutions (Shankari 1991; Janakarajan 1993). This decline has greatly accelerated since the 1980s, wherein the individualised sinking of borewells tended to undercut the social compacts integral to tank maintenance and operation, leading to the latter falling into disrepair (Balasubramanian 2008). Over the past two decades, the area of land under tank irrigation has not simply declined in relation to canal and well irrigation but has undergone an absolute reduction across semi-arid Andhra Pradesh. In the Telangana region there was a 58 per cent decrease in tank-irrigated area between 1956–57 and 2005–09. In Rayalaseema, the corresponding period saw a 70 per cent decrease in tank-irrigated acreage (Pingle 2011; see also, Sreedhar 2007). Notably, because tanks play an important role in recharging groundwater levels, the decline of tank systems creates a negative feedback loop into aquifer depletion that amplifies the speed of the latter over the course of the season.

Second, the exploitation of groundwater has reached an upper threshold in many semi-arid regions. Since the early 2000s, Andhra Pradesh as a whole has seen decreasing returns from groundwater irrigation and the escalating prevalence of well failure. The special secretary to the state's irrigation development department noted that the quantum of groundwater Andhra Pradesh received earlier from 100,000 borewells is now obtained by 260,000 borewells (cited in Mishra 2010). Similarly, Kumar et al. (2011) note that farmers invested Rs 1,648 crore (approx. $300 million) for drilling new wells between 2000 and 2004 without adding any extra area irrigated for their efforts. By abstracting water from the same aquifers, farmers were effectively redistributing existing water among a greater number of wells. In short, the proliferation of wells was further dividing a relatively static and overexploited resource, leading to diminished yields to each well. The same authors thereby conclude that groundwater irrigation had effectively reached a saturation point by 2000–01, wherein the groundwater irrigation potential was sufficient to harness all the renewable groundwater resources that occur in a good rainfall year (Kumar et al. 2011: 12).

As a consequence, groundwater depletion is rampant, particularly following years of below-average rainfall. Government estimates for Andhra Pradesh in 2008 suggested that 41 per cent of groundwater blocks were at overexploited, critical or semi-critical levels. Such aggregated state-level totals, however, do not distinguish for regional variations. In particular the manifold differences between the conjunctive use of groundwater in areas already served by canals and that occurring outside of such zones. For instance, in the drought-prone Anantapur district, southwest of Andhra Pradesh, out of a total 63 sub-districts, more than half (33) were judged to be overexploited, placing them in what is termed the 'dark zone' of acute resource depletion wherein more water is abstracted than is replenished over the course of the year. A further five were in a critical state (90–100 per cent utilisation of water vis-à-vis recharge), 13 semi-critical (70 to 90 per cent utilisation) and only 12 were 'safe' (below 70 per cent) (Rukmani and Manjula 2010). Numerous analysts also contended that official methods of counting groundwater recharge and depletion tend to overestimate the quantities of water available, thereby strongly underestimating the level of aquifer depletion across the state (Singh and Singh 2002; World Bank 2010: 62; Kumar et al. 2011).

This record suggests that strong socio-ecological limits to the groundwater revolution in rain-fed areas are being manifested in the form of overexploitation of aquifers and declining water tables. This crisis of groundwater overexploitation has not gone unnoticed by either the Andhra Pradesh state government or the World Bank, both of which have heavily promoted extraction as part of agrarian modernisation policies. Many critics have noted how the free electricity policies promote inefficiency (Kondepati 2011). Electrical pumps are extremely energy intensive, and such energy use escalates the deeper the pumps extend. Given the uncertainty of when electricity provision will occur, farmers leave the pumps constantly turned on waiting for the irregularly delivered electricity to arrive. This not only promotes wastage but also increases the incidence of pump burn out, adding further costs to farmers. At the same time, electricity subsidies provide a significant burden upon state budgets, accounting for around 50 per cent of the annual fiscal deficit. Regardless of such issues, the promise of free electricity to rural smallholders has become politically sacrosanct. An attempt to partially withdraw subsidies in Andhra Pradesh in 2000 led to violent protests by farmers mindful of their keen dependence on groundwater. The subsidies were soon reinstated and then transformed in 2006 into the policy of free electricity provision under the Congress administration.

In this context, the Andhra government has maintained subsidised electricity while emphasising regulatory options for curbing groundwater overexploitation. The 2002 Andhra Pradesh Land Water and Trees Act promulgated a new framework that created a specific state body to oversee the management of groundwater resources. In particular, the act allowed the state to curtail the drilling of new wells in over-exploited districts and to insist on the registration of new wells in all districts and the appropriate spacing of those wells. As an incentive for farmers to comply with the act, registered wells could be enrolled in a government insurance scheme that would compensate farmers should the initial drilling fail to encounter water. The results, however, have been underwhelming, with implementation of the act in rural areas partial at best. For example, while in 2005–6 the number of formally registered wells showed a clear decrease from previous years, the number of electricity connections for wells increased far in excess of well permits, indicating that there is a general disregard for the new laws and illegal well drilling is commonplace (Rama Mohan 2008; Mishra 2010). Figures from the semi-arid regions of Telangana similarly show the number of connections increasing, with a 13 per cent jump between 2004 and 2009, while the connected load increased by 26 per cent over the same period, again indicating the widespread non-registering of wells (Pingle 2011).

The failures of regulation to significantly impact upon groundwater abstraction has prompted the World Bank, a major promoter of groundwater development, to encourage community-based water management as a solution to the ills of overexploitation in Andhra Pradesh and elsewhere in India (World Bank 2010). Several projects that were begun by aid agencies and NGOs in the late 1980s were expanded and formalised in the 1990s and 2000s. The specific model that has been most widely promoted over the past decade is known as the Andhra Pradesh Farmer Managed Groundwater Systems (APFMGS). The latter, spearheaded by the World Bank and the Food and Agriculture Organisation and supported by the state government, promotes an NGO-driven process wherein water-management experts train local communities to monitor rainfall and groundwater levels so as to better understand the dynamics of extraction and recharge in their region. These experts subsequently provide a guideline for community cropping strategies based on projected groundwater levels for the season ahead. This frequently involves, for example, recommending a switch from water-intensive rice paddy to alternate crops should water levels decrease below a certain level in the lead up to planting season.

The APFMGS approach is grounded in a framework that conceives of groundwater overexploitation as produced primarily by information deficiencies that prohibit farmers from making rational decisions about which crops to plant. By educating farmers on the causes of groundwater depletion and training them to collectively monitor groundwater levels, the World Bank projects the model as an effective tool to counter over-extraction based on voluntary and rational self-discipline. As the World Bank terms it:

> [T]he project does not seek collective action on reducing groundwater abstraction, and individual farmers are free to plant what they want and pump as they desire. The reductions in groundwater draft in APFAMGS are not coming from altruistic collective action, but from the individual

> risk management and profit-seeking decisions of thousands of farmers. This makes the APFAMGS model robust and replicable, as no authoritative leadership is required for enforcement of compacts.
>
> (World Bank 2010: xv)

The recourse to overcoming information asymmetries that prohibit rational action fits well within the bank's prevailing neoinstitutionalist framework discussed in Chapter Four. The degree to which the programme can be successful, however, is yet to be ascertained. For the World Bank, initial studies suggest that the programme is a success, with test communities changing cropping patterns according to water levels, leading to a more efficient use of groundwater and better yields. Notably, however, the comparison years chosen for the bank's study were 2005 – a drought year that followed a relatively dry period – and 2008 – a year of relatively plentiful rainfall that followed two years of average rainfall. Whether any strong conclusions can be drawn from such distinct hydroclimatic periods between which aquifer levels greatly increased is unclear. The real test of the community-driven approach is likely to occur under periods of below-average rainfall, wherein the pressures of crop failure induce significant livelihood stresses and exacerbate conflict over resource use. In this respect, a different review of the project notes that, despite early successes, there remain fears that once the NGOs and state organisations withdraw, self-organising activities tend to decline. In the case highlighted by the NGO Action for Food Production, the end of direct project organising led to a return of relatively indiscriminate borewell drilling culminating in the overexploitation of groundwater resources that such projects were anticipated to prevent (AFPRO 2006: 55).

This raises the question of why supposedly rational cultivators would continue to pursue a collectively irrational form of action despite common knowledge of the risks involved. Two factors need to be highlighted that are obscured in the APFMGS approach. First, the ways in which control over water are interwoven into hierarchies of caste and class are entirely exorcised from the World Bank's literature on community-managed systems. While the bank frames water use in a paradigm of rational and voluntaristic individual decision-making, community-based groundwater management cannot be divorced from the social relations of power that stratify agrarian societies. In the context of deeply embedded social hierarchies, forms of class and caste power profoundly shape the parameters of what constitutes 'rational' actions for dominant and dominated groups. Notably, the Independent Evaluation Group (IEG) that monitors World Bank programmes professed considerable concern about the role of power dynamics within water user associations for tank and canal systems within Andhra Pradesh. For example, the IEG (2008: 23) highlighted how a survey of 222 canal system water user associations (WUA) in 22 districts revealed that 88 per cent of leaders were drawn from higher castes and 11 per cent from backward castes, with just one per cent from scheduled tribes or scheduled castes. Despite these initiatives being labelled as a 'pro-poor' participatory irrigation management strategy, large landowners accounted for 88 per cent of WUA presidents. Similarly, on the basis of an ethnographic study of the formation of watershed development committees in Kurnool district, Vasudha Chhotray argues that the formation of such bodies are deeply mired within relationships of power and domination that stratify village politics and which unevenly distribute the gains and costs of such programmes between classes and castes (Chhotray 2011: p. 150). While the class and caste dimensions of community groundwater management bodies are not yet clear, these previous experiences suggest that the emphasis on community self-organisation can strongly reinforce existing social hierarchies.

Similarly, there is also a conspicuous silence surrounding the differential access to credit and the social relations of indebtedness that are central to both well drilling in particular and agricultural production in general. For example, in its long assessment of groundwater crisis in India, the World Bank mentions debt only once, as an outcome of well failure rather than a precondition of well drilling (World Bank 2010). In such accounts, the effects of debt in shaping individual and collective actions are entirely marginalised despite their centrality to shaping household decision-making. As I draw out below, these dynamics are essential for understanding the possibilities and limits to community-based management because it leads to a quite different understanding of the motivations involved as compared to the rational-actor paradigm. In the context of debt-driven agriculture, smallholders are not drilling borewells and over-extracting water as a voluntary livelihood strategy measured against the risk and costs of various possible courses of action. Rather, they are drilling borewells under the pressure of constant debt cycles, under which the weight of crop failure is immense and the option to scale back production may not be viable despite it being 'rational' given projected groundwater levels. To flesh out these socio-ecological dynamics, the following section contextualises groundwater overexploitation within the transformation of the political ecology of agrarian relations in semi-arid Andhra Pradesh ongoing over the past three decades.

The political ecology of debt-driven agriculture

As both macro- and village-level studies have demonstrated, agrarian Andhra Pradesh has undergone a significant shift towards increasingly commercialised agriculture on both larger farms and for smallholders (Niranjan Rao and Narashima Reddy 2008; Ramachandran, Rawal and Swaminathan 2010). A changed macro-policy framework operating since the early 1990s, in which input subsidies and tariffs were reduced and other liberalisation measures undertaken, has accentuated this transformation. Under the rubric of promoting greater efficiency and productivity in the agricultural sector, these reforms were envisaged to promote a rationalisation of the agrarian economy in ways that could improve aggregate crop yields and aid the expansion of the off-farm economy. These policy changes, however, were underwhelming in their stated aim of promoting agricultural growth and came at a significant cost for large sections of the rural population. On the one hand, agricultural growth slowed over the 1995–2005 period compared to the previous decade. On the other, the benefits of growth were highly uneven, with larger farmers accounting for a greater share of output gains, while smallholders faced acute pressures on their ability to maintain agriculture-based livelihoods (Vakulabharanam and Motiram 2011). As such, these policy reforms served to accentuate existing trends towards social differentiation among the agrarian population across semi-arid areas of India in general and Andhra Pradesh in particular (Harriss-White 2008). For smallholders, the reduction of subsidies on key inputs allied to retrenchment of protective tariffs placed huge strains on their ability to turn a profit from agricultural production (Reddy and Mishra 2009). Equally, the scaling back of the social banking initiatives of the 1970s and 1980s tended to restrict the supply of credit, creating increased dependence on rural moneylenders and opening new avenues for surplus extraction between social classes (Taylor 2012).

At the same time, demographic expansion hastened the subdivision of already small landholdings. In the context of an existing legacy of unequal land distribution, population increase created a mass of rural households for whom land constraints are a significant barrier to household reproduction through agriculture. At an all-India level, by 2005–06 some 65 per cent of agricultural households owned less than one hectare (marginal) and a further 19 per cent owned between one and two hectares (small) (Basole and Basu 2011). The intense pressures upon households with such smallholdings to provide the means for subsistence resulted in them incorporating diverse livelihood strategies to make ends meet. There is a particular emphasis on migration, with increasing numbers of circular migrants who travel greater distances in search of waged employment (Deshingkar et al. 2009: 63). These shifting flows of labouring bodies across the agrarian environment have created a new cartography of social reproduction across rural Andhra Pradesh. Landless, small and marginal farmers in particular are increasingly reliant on forms of circular migration to supplement agricultural earnings and to pay off debts from agriculture and other expenditures. Notably, while men increasingly tend to migrate for seasonal employment, women have been drawn in greater numbers into agricultural wage-labour. This has fashioned a new gendered division of labour in the countryside, marked by notable inequalities in which women's wages for agricultural work are comprehensively below those paid to men (Garikipati 2008).

While the increasing prevalence of migration indicates that agriculture is no longer a sufficient means of social reproduction for a majority of households, the opportunities for off-farm employment nevertheless have expanded in an uneven fashion over the past two decades. Throughout the 1990s and early 2000s, off-farm employment remained stagnant while the margins for agricultural production tightened (Centre for Economic and Social Studies 2008). Livelihood 'diversification' has proved extremely uneven. A detailed comparative village level study from Telangana in northwest Andhra, for example, noted that while some diversification strategies did succeed in increasing real incomes for some, for most they offered "little more than an opportunity to cope and mitigate risk or to tread water and hold on to productive assets for the future" (Kumar Deb et al. 2002: 33). Enduring ties to agriculture therein remain a backbone of many rural livelihoods, and smallholders remain compelled to find ways to maximise the return from their limited lands alongside pursuing livelihood diversification. To address this structural vulnerability, households have increasingly self-exploited by using greater amounts of unpaid household labour for agricultural work alongside increasing risk-taking through debt-leveraged cropping strategies in order to try and expand incomes.

This has led to a transformation of the social, ecological and technological parameters of smallholder agriculture. New cropping techniques have changed the social organisation of production, including the demand for labour, the nature and timing of labour activities and the broader integration of production into local, national and international markets. In the pursuit of increasingly specialised production, smallholders have been incorporated into the commercialised provision of seeds and other inputs – particularly pesticides and fertiliser – from a range of public and private bodies. A study by the Society for Elimination of Rural Poverty, for example, noted that smallholders in Andhra spent around 35 per cent of their total agricultural outlay on fertiliser and pesticides alone (Kumar et al. 2009). This has imposed a dynamic towards a greater intensity of cropping, including the increased use of chemical inputs and commercial seeds. On this basis, smallholder production has become increasingly productive by allaying itself of technology and irrigation in a fraught attempt to maintain profitable agricultural production in increasingly competitive conditions. One important study has detailed how smallholder agriculture in India is more productive in terms of yield per hectare than larger farms and uses chemical inputs more intensively (Chand, Prasanna and Singh 2011). In comparison to large-size holdings (>4 ha), marginal (<1 ha) and small (1–2 ha) farms used 47 and 20 per cent more fertilizer, respectively, per unit of irrigated land and 336 and 245 per cent more on un-irrigated land. By 2001, the use of HYV seeds was 72 and 58 per cent on marginal and small farms respectively but below 50 per cent on large farms (Chand, Prasanna and Singh 2011).

Relationships of credit/debt and access to irrigation are evidently key mediators of this process. As noted above, capitalised agriculture requires both regular flows of water and power and intensive market-purchased inputs for which the costs have increased markedly as government subsidies were retrenched over the neoliberal period. As a consequence, at the beginning of the growing season, many inputs are purchased on credit stemming from local merchants, moneylenders, seed companies, agricultural cooperatives or through overlapping microfinance loans. As Chapter Four noted, these sources of credit involve entry into multiple contracts in which landlords or merchants tie loans to agreements for the purchase of the producer's crop at a time and price most advantageous to them. Indebtedness is rampant across rural Andhra Pradesh, with 80 per cent of farmers holding debts and virtually all small and marginal farmers heavily indebted to both formal and informal sources (Kumar et al. 2009; see Ramachandran, Rawal and Swaminathan 2010 for village-level surveys). The need to cover forthcoming debt repayments leads farmers to gravitate further towards cash crops such as chillies, sugarcane and cotton alongside rice, with the latter favoured by government procurement policies. This shift towards the monocropping of paddy and cash crops has displaced traditional crops such as sorghum (*jonnalu*), finger millet (*ragi*) and pearl millet (*sajjalu*) that require less water, can grow in less fertile soil, offer greater heat resistance and enjoy shorter growing cycles. Concurrently, monocropping and the use of transgenic seeds reinforce an escalating dependence on technological inputs and the necessity of reliable water in a semi-arid climate.

Attempting to exercise control over flows of water, therefore, is central to this struggle to maintain household reproduction within an austere agrarian environment. In some cases, the entire *kharif* (monsoon) crop can be sold simply to pay off debts, leaving smallholders dependent on the *rabi* (dry season) crop and supplemental sources of income to facilitate household reproduction. Outside of areas sourced by reliable canal irrigation, the *rabi* crop is largely dependent upon borewell irrigation despite the ecological uncertainties of drilling into hard-rock aquifers. In contrast to the model of rational household livelihood strategies used by the World Bank, it is precisely these debt-driven processes that underscore the proliferation of well drilling despite the increasing risks of failure and the decreasing returns of fluid to each well. As one groundnut farmer in Andhra Pradesh, who has two failed borewells and a debt of Rs 1 lakh with an interest of three per cent per month, noted, "We need water and cannot grow anything without borewells, so we keep trying despite failures" (cited in Mishra 2010: 2). For smallholders, there is a cycle of overlapping risks. Accessing groundwater might succeed in pushing back ecological barriers to production by providing more regular water, but it replaces those constraints with the social relations of debt, a new temporal horizon of repayment and the grave risk of failed wells and intractable debts.

Despite being overlooked in the policy literature described above, the class dimensions of these processes are central to the uneven distribution of these risks.[3] Notwithstanding claims to the 'democratisation of irrigation', access to groundwater remains conditioned by a household's ownership of land and their ability to afford renting a drilling crew, purchasing a pumpset, securing an electrical connection and undertaking ongoing repairs. These capital costs are met either through profits, bank loans or – most commonly for smallholders – by drawing upon informal sources of credit. Notably, over the 1990s state credit for smallholder agriculture was firmly retrenched and all indices of banking inclusiveness declined (Shetty 2009). While in the later 2000s the Indian government unleashed much-heralded policy targets for banks to reinvest in agriculture, banks have tended to meet these targets in ways that benefited large agribusiness over small or marginal farmers. As a result, the increase in agricultural lending over the past ten years has been unevenly spread, creating a bifurcation between bank financing that benefits larger, capitalist farmers and a microfinance policy that injects small infusions of credit at a household level through the medium of rural women. For example, almost half the total 'agricultural' credit in the state of Maharashtra in 2008 was disbursed by urban and metro bank branches, more than 42 per cent of it in Mumbai alone which, as noted journalist P. Sainath puts it, "is the stomping ground of large corporations rather than of small farmers" (2010b).

For marginal and smallholder farmers, a lack of collateral allied to the widely perceived insecurity of agricultural returns mean that accessing institutional credit for agricultural inputs remains extremely difficult (Mehrotra 2011). Simultaneously, microcredit, which is aggressively promoted in rural India, is precisely too micro to cover the costs of well drilling plus the pump and pipe equipment, which range between IR 30,000 and 80,000 (approx. $550–$1,500) depending on the depth of the well. As a consequence, informal loans at exploitative rates of interest remain the primary option for many smallholders despite locking them into relations of surplus extraction and dependency that are hard to break. The potential for debt traps is written into this political ecology of agricultural production. If drilling is successful and the well taps into an aquifer that remains functional throughout the season, then costs could be recovered through increased yields within a one to four year time-span (AFPRO 2006: 34). Those two prerequisites for success, however, are increasingly suspect under conditions of generalised aquifer depletion. Rama Mohan (2012), in his fieldwork in semi-arid northwest Andhra Pradesh, for example, suggests that on average for each successfully drilled borewell that distributes water for part of the year, there are two complete failures.

Consequently, while direct access to groundwater appears to smallholders as an immediate ecological fix to the pressures of competitive agriculture on limited land holdings, it comes at the cost of deepening the structural pressures of debt and the risks of catastrophic failure. In areas of high groundwater abstraction, there is a vicious cycle ongoing. Competitive drilling and aquifer depletion over the course of a season lowers the water level, making earlier wells obsolete and compelling a new round of deeper drilling. This process manifests itself first with more marginal wells that are either dug more shallowly or extend into the edge of the aquifer. As the dry season moves on, however, increasing rates of well failure are recorded. A field study in the Medak region in the northwest of the state, for example, revealed that that 244 out of 250 borewells that were functional during and immediately after the monsoons of 2009 went partially or fully dry in subsequent summer months irrespective of the stage of groundwater development – i.e. 'safe' or 'critical' (Rama Mohan 2012). In short, while an aquifer might still technically contain water, as the water table decreases this causes failure for numerous wells that are either sub-optimally placed above the aquifer or that are less deeply dug.

As such, more affluent farmers with the credit resources to continually deepen their wells are able to undercut their neighbours in the race to the bottom of the water table. This has marked consequences for those who have gone into debt to drill wells only to find them subsequently drying up early in the season. The upshot is that smallholders are increasingly pressured not only to drill their own wells but to continually deepen them or drill in alternate locations when existing wells run dry, therein deepening their indebtedness. The results can be devastating. The Andhra Pradesh Human Development Report of 2007, for example, noted the endemic levels of well failure in Musapally village in Nalgonda district where, for a village with 2,000 acres under cultivation, there are more borewells than people. In this location, some 6,000 borewells were drilled at an estimated investment of Rs. 6.52 crore (approx. $1.2 million) and about 85 per cent of these wells failed (Centre for Economic and Social Studies 2008: 72). While the creation of groundwater markets – wherein those with working borewells sell irrigation by the hour to neighbouring fields – was expected to overcome these problems, studies in Andhra Pradesh and elsewhere indicate that, first, there is insufficient water available to meet demand and, second, control over such flows allows providers to reap a considerable surplus from those without access.[4]

Given the dependence of smallholders upon debt, credit markets form a significant means of surplus extraction between agrarian classes. With the risks involved in agricultural production, engaging in money lending and other extractive forms of accumulation is a key source of income for social classes that hold capital.[5] For those who depend at least partially upon agriculture, however, the barriers posed by groundwater depletion express the socio-ecological contradictions of smallholder livelihoods. While tapping into groundwater initially provided a windfall for those with the resources to do so, the risks involved in drilling have subsequently expanded alongside the proliferation of wells. It is on this basis that there are strong links between cashcropping, the failure of borewells, overwhelming debts and farmer suicides in the semi-arid regions of the Deccan plateau (Galab, Revathi and Reddy 2009; Mishra 2010; Reddy 2010; Vasavi 2012). A study of 30 farmer suicides that occurred in the mid-2000s in Anantapur district of Andhra Pradesh, for example, described how heavy indebtedness owing to agricultural credit from primarily informal sources was the underlying factor in the vast majority of cases. Specifically, the failure of borewells was directly correlated to 30 per cent of these suicides, while another 40 per cent related to the medium term build up of agricultural debts, including irrigation expenses from drilling wells or buying groundwater. As the authors note, such farmers had "pinned their hopes almost exclusively on the anticipated income from agriculture to repay their dues" (Sreedhar 2010: 229). In these tragic cases, the lifeline of credit ran into the temporal wall of insurmountable debts, to which suicide appeared a sole means of escape.

Climatic change and the politics of the agrarian environment

How then, might we situate the notion of 'climate change adaptation' in this kind of socio-ecological landscape wherein the insecurity of households is embedded in a broad spectrum of socio-ecological relations operating across scales that condition household social reproduction? As is clear, projected climatic change in the form of increasingly erratic rainfall threatens to further tighten the constraints upon marginal farmers and smallholders. This is an expressly worrying scenario. Such trends, however, are only comprehensible as part of the socio-ecological matrix through which the agrarian environment is produced as a whole, including the way that debt relations are messily entangled with control over labouring bodies and flows of water. In this respect, climatic change will be expressed not simply in changing rainfall patterns but simultaneously and inextricably through the ongoing transformation of social relations. This process will continue to open opportunities for accumulation to those social classes that are able to exercise greater control over key elements within the agrarian environment, such as water, land and capital, not to mention institutionalised political power. These trends, of course, are not independent but are related moments of this dynamic production of the lived environment. Fundamentally, they point to how working with climatic change cannot in any meaningful way be separated out from a redistribution of power within the agrarian environment.

Bringing the socio-ecology of power into the questions imposed upon us by climatic change militates against the mainstream conceptual apparatus. As noted above, in the face of these socio-ecological dynamics, the World Bank deploys a theoretical frame that portrays groundwater use as a voluntary action undertaken by rational utility-maximising farmers who operate in conditions of incomplete information vis-à-vis an external environmental constraint. This framing of rural social relations was deployed to legitimise a form of community organising predicated upon groundwater overexploitation being caused by farmers who are unable to read the water table, thereby rendering them incapable of planning rational cropping patterns. On the contrary, the presence of engrained power relationships and the enduring burden of debt undermines the attempt to project the decisions of smallholders as simply the outcome of rational utility maximisation in conditions of imperfect information. While the strategies of community-based groundwater management predicated on promoting 'water literacy' may indeed provide farmers with useful information, they are unlikely to stem the cycle of groundwater depletion to the extent that it is driven by debt-driven competition in an austere agrarian environment. Specifically, the ability of more affluent farmers to dig deeper wells and therein undercut their neighbours in conditions of groundwater scarcity emphasises how these socio-ecological dynamics of agrarian distress cannot be understood in abstraction from power over land, water and credit. This betrays the fallacy of purely localistic solutions to issues that are determined by forces acting on multiple and overlapping scales.

While this emphasis on power and scale might be contentious among some scholars of climate change adaptation, the point is clearly understood across the region. Andhra Pradesh, particularly in the northern region of Telangana, has been a centre of violent agrarian conflicts that targeted the embedded power of landlords to exercise control over land, labour and water. Such struggles reached their apex in the 1970s, in which state repression quelled militancy although not before many of the more feudalistic forms of agrarian power had been fractured (Singharoy 2004). In more recent times, political agitation has centred upon the bifurcation of the state and the separation of Telangana from Andhra. This political movement has – at the time of writing – seemingly been successful with the Congress government proclaiming the imminent breakup of the state, although the political wrangling over the details continues. Whether independent statehood for Telangana can adequately address the longstanding questions of control over water flows, agrarian investment and the employment opportunities provided through the institutions of the state is unclear. There is perhaps a justified fear that the political aspirations that underscore the movement may be co-opted within an agrarian populism that simply maintains or increases the electricity subsidy for borewells (see Swain and Charnoz 2012 for discussion). Under such conditions, expanded groundwater exploitation will continue to be seen by many smallholders as the primary strategy by which to overcome the impacts posed by climatic change on rainfall patterns, notwithstanding its socio-ecological contradictions and its compounding effect as a carbon-intensive form of irrigation.[6]

Alongside such political agitation, however, other incipient transformations are noteworthy. Several areas of the semi-arid Deccan have witnessed a pushback against Green Revolution technologies and a partial transition, in some regions, towards agro-ecology orientated forms of production. Given the scale of problems raised by capitalised smallholder agriculture, farmers movements and rural development societies have repeatedly emphasised how a revalidation of diverse cropping strategies, including the use of biodiverse farming systems, can alleviate pressure on water resources and better ensure food security within community-driven agriculture.[7] Notably, a World Bank-sponsored report based on field research in Andhra indicated how the intensive application of chemical inputs, while creating the grounds for escalating debt traps, provides no yield gains compared to integrated community organic farming methods (Kumar et al. 2009).

Agro-ecological techniques, however, can only be successful if conceptualised and implemented within a broader frame of socio-ecological relations that expand far beyond the scale of the farm (van der Ploeg 2008). In this respect, rural social movements in the region have repeatedly articulated a much broader set of transformations that could collectively help alter the socio-ecological basis for livelihoods through the region. As noted in Chapter Four, these include substantive land reform, an end to commercial encroachment on common property resources, protection of agriculture from foreign imports, a restructuring of rights over irrigation, effective guaranteed minimum prices for crops, protection of labourers from the power of landlords and labour contractors and subsidised forms of credit made available to smallholders and tenant farmers (Commission on Farmers' Welfare 2004). It is notable that such demands take aim at the underlying determinants of power within the agrarian environment rather than solely a more equitable distribution of market opportunities and targeted public policies. The question may well become to what extent the language of climate change adaptation either facilitates or hinders the expression of these longstanding issues of building more equitable and sustainable agrarian environments.

Notes

1 It is not surprising, therefore, that the Ministry of Environment and Forests instigated a State Action Plan on Climate Change drawn up by the Environment Protection Training and Research Institute, based in the state capital of Hyderabad (EPTRI 2011). As is habitual within the discourse of climate change adaptation, the plan conceptualises socio-ecological change across the region through the prism of vulnerability, adaptive capacity and resilience. Indeed, the plan frames the need for adaptation to counter

vulnerability through minimising the exposure of social and ecological 'systems' and increasing their adaptive capacity.

2 Many small and marginal farmers instead turn to local diviners, who use traditional methods involving sticks or coconut shells at a fraction of the cost yet with uneven records of success.

3 S. Janakarajan has detailed at length the class dynamics of groundwater access in his work on Tamil Nadu (e.g. Janakarajan 2002; Janakarajan and Moench 2006)

4 A study in across villages in Andhra suggested that not only did non-pump owners pay 'extortionate' prices per hour of water, they were also unable to purchase the requisite amount needed for the irrigation of their crops (Adiseshu and Madhu Babu 2007).

5 Despite the increasing penetration of microfinance into rural Andhra Pradesh over the past two decades, informal credit markets have undergone a parallel expansion over the same period (Taylor 2011).

6 For each extra meter of decline in aquifer depth, the increase in pumping necessary to extract water raises emissions by six per cent. Groundwater abstraction is therein estimated to account for 4–6 per cent – and rising – of India's total carbon emissions (Shah 2009).

7 For example, see the work of the Deccan Development Society (Deccan Development Society 2009).

References

Adiseshu, K., and K. Madhu Babu. 2007. "Groundwater Market Dynamics in the Villages of Andhra Pradesh." Pp. 210–27 in *Irrigation Development: Issues and Challenges*, edited by P Narashima Rao. New Delhi: Discovery.

AFPRO (Action for Food Production). 2006. *A Systematic Assessment of Community Based Groundwater Management Experiences in Andhra Pradesh*. Hyderabad: Irrigation & Command Area Development Department, Government of Andhra Pradesh.

Arnold, D., and R. Guha. 1995. *Nature, Culture, Imperialism: Essays on the Environmental History of South Asia*. Delhi: Oxford University Press.

Balasubramanian, R. 2008. "Community Tanks vs. Private Wells: Coping Strategies and Sustainability Issues in South India." Pp. 283–305 in *Promise, Trust and Evolution: Managing the Commons of South Asia*, edited by Rucha Ghate, Narpat Jodha, and Pranab Mukhopadhyay. New Delhi: Oxford University Press.

Banaji, J. 1977. "Capitalist domination and the small peasantry: Deccan districts in the late nineteenth century." *Economic & Political Weekly* 12(33–4): 1375–404.

Basole, A., and D. Basu. 2011. "Relations of production and modes of surplus extraction in India: part I – agriculture." *Economic & Political Weekly* XLVI(14): 41–59.

Baviskar, A. 2007. *Waterscapes: The Cultural Politics of a Natural Resource*. New Delhi: Permanent Black.

Centre for Economic and Social Studies. 2008. *Human Development Report 2007: Andhra Pradesh*. Hyderabad: Government of Andhra Pradesh.

Chand, R., L. Prasanna, and A. Singh. 2011. "Farm Size and Productivity: Understanding the Strengths of Smallholders and Improving Their Livelihoods." *Economic & Political Weekly* XLVI(26 & 27): 5–14.

Chhotray, V. 2011. *The Anti-Politics Machine in India: State, Decentralization, and Participatory Watershed Development*. London; New York: Anthem Press.

Commission on Farmers' Welfare. 2004. "Commission on Farmer's Welfare Final Report." Hyderabad: Government of Andhra Pradesh.

Deccan Development Society. 2009. *Agrobiodiversity Heritage Site of Zaheerabad, Medak District, Andhra Pradesh: A Profile*. Medak, Andhra Pradesh: Deccan Development Society.

Deshingkar, P., L. Rao, S. Akter, and J. Farrington. 2009. "The Evolving Pattern of Circular Migration: Household Surveys in Andhra Pradesh." Pp. 58–87 in *Circular Migration and Multilocational Livelihood Strategies in Rural India*, edited by Priya Deshingkar and John Farrington. New Delhi: Oxford University Press.

EPTRI (Environment Protection Training and Research Institute). 2011. "State Action Plan on Climate Change for Andhra Pradesh." Edited by Ministry of Environment and Forests. Hyderabad: EPTRI.

Gadgil, M., and R. Guha. 1993. *This Fissured Land: An Ecological History of India*. Berkeley: University of California Press.

Gadgil, M., and R. Guha. 1994. "Ecological conflicts and the environmental movement in India." *Development and Change* 25(1): 101–36.

Galab, S., E. Revathi, and P. P. Reddy. 2009. "Farmers' Suicides and Unfolding Agrarian Crisis in Andhra Pradesh." Pp. 164–98 in *Agrarian Crisis in India*, edited by D. Narasimha Reddy and Srijit Mishra. New Delhi: Oxford University Press.

Garikipati, S. 2008. "Agricultural wage work, seasonal migration and the widening gender gap: evidence from a semi-arid region of Andhra Pradesh." *European Journal of Development Research* 20(4): 629–48.

Guérin, I., M. Roesch, G. Venkatasubramanian, and S. Kumar. 2011. "The social meaning of over-indebtedness and creditworthiness in the context of poor rural South India households (Tamil Nadu)." *RUME Working Papers Series* (2011–1).

Harriss-White, B. 2008. "India's Rainfed Agricultural Dystopia." *European Journal of Development Research* 20(4): 549–61.

Hill, C. 2008. *South Asia: An Environmental History*. Santa Barbara, CA: ABC-CLIO.

IEG (Independent Evaluation Group). 2008. *An Impact Evaluation of India's Second and Third Andhra Pradesh Irrigation Projects: A Case of Poverty Reduction with Low Economic Returns*. Washington, DC: IBRD.

Janakarajan, S. 1993. "In search of tanks: Some hidden facts." *Economic & Political Weekly* 28(26): 53–60.

Janakarajan, S. 2002. "Wells and Illfare: An Overview of Groundwater Use and Abuse in Tamil Nadu, South India." In *IWMI-Tata Water Policy Research Program Annual Meeting, 2002*. Mumbai: IWMI-TATA Water Policy Program.

Janakarajan, S., and M. Moench. 2006. "Are wells a potential threat to farmer's well-being? Case of deteriorating groundwater irrigation in Tamil Nadu." *Economic & Political Weekly* 41(37): 3977–87.

Kondepati, R. 2011. "Agricultural groundwater management in Andhra Pradesh, India: A focus on free electricity policy and its reform." *International Journal of Water Resources Development* 27(2): 375–86.

Kumar Deb, U., G. N. Rao, Y. Mohan Rao, and R. Slater. 2002. *Diversification and Livelihood Options: A Study of Two Villages in Andhra Pradesh, India, 1975–2001*. Working Paper 178. London: Overseas Development Institute.

Kumar, M. D., M. Sivamohan, V. Niranjan, and N. Bassi. 2011. "Groundwater Management in Andhra Pradesh: Time to Address Real Issues." Hyderabad: Institute for Resource Analysis and Policy.

Kumar, T. V., D. V. Raidu, J. Killi, M. Pillai, P. Shah, V. Kalavakonda, and S. Lakhey. 2009. *Ecologically Sound, Economically Viable: Community Managed Sustainable Agriculture in Andhra Pradesh, India*. Washington, DC: World Bank.

Lambrou, Y., and S. Nelson. 2010. *Farmers in a Changing Climate – Does Gender Matter? Food Security in Andhra Pradesh, India*. Rome: Food and Agriculture Organization.

Ludden, D. 1999. *An Agrarian History of South Asia*. Cambridge: Cambridge University Press.

Mehrotra, N. 2011. "Agriculture credit: The truth behind the aggregate numbers." *Economic & Political Weekly* XLVI(42): 22–28.

Mishra, A. 2010, July 3. "Andhra's groundwater crisis and a trail of suicides." *Down to Earth*. Available at: http://www.downtoearth.org.in/content/andhras-groundwater-crisis-trail-suicides

Mosse, D. 2003. *The Rule of Water: Statecraft, Ecology and Collective Action in South India*. Oxford: Oxford University Press.

Mukherji, A., and T. Shah. 2005. "Groundwater socio-ecology and governance: a review of institutions and policies in selected countries." *Hydrogeology Journal* 13(1): 328–45.

Nag, K. 2011. *Battleground Telangana: Chronicle of an Agitation*. New Delhi: Harper Collins.

Niranjan Rao, G., and D. Narashima Reddy (Eds.). 2008. *Rural Transformation: Perspectives from Village Studies in Andhra Pradesh*. Delhi: Daanish Press.

Olsen, W. K. 1996. *Rural Indian Social Relation: A Study of Southern Andhra Pradesh*. Delhi: Oxford University Press.

Pande, P., and K. Akermann. 2010. *Adaptation of Small Scale Farmers to Climatic Risks in India*. New Delhi: Sustainet India.

Pingle, G. 2011. "Irrigation in Telangana: The rise and fall of tanks." *Economic & Political Weekly* XLVI(26 & 27): 123–31.

Rama Mohan, R. 2012. "Failure of Agricultural Bore Wells in Hardrock Areas of Andhra Pradesh: A Diagnostic Analysis." Secunderabad: European Union/Centre for World Solidarity.

Rama Mohan, V. R. 2008. "Groundwater in Andhra Pradesh, India: The Case of Privatization of a Common." In *International Association for the Study of Commons Conference 2008: Governing Shared Resources*, University of Gloucestershire.

Ramachandran, V. K., V. Rawal, and M. Swaminathan. 2010. *Socio-economic Surveys of Three Villages in Andhra Pradesh: A Study of Agrarian Relations*. New Delhi: Tulika Books.

Reddy, D. N., and S. Mishra. 2009. "Agriculture in the Reforms Regime." Pp. 4–43 in *Agrarian Crisis in India*, edited by D. Narasimha Reddy and Srijit Mishra. New Delhi: Oxford University Press.

Reddy, T. P. 2010. "Distress and Deceased in Andhra Pradesh: An Analysis of Causes of Farmer' Suicide." Pp. 242–63 in *Agrarian Crisis and Farmer Suicides*, edited by R.S. Deshpande and S. Arora. New Delhi: Sage.

Rukmani, R., and M. Manjula. 2010. *Designing Rural Technology Delivery Systems for Mitigating Agricultural Distress: A Study of Anantapur District*. Chennai, India: M.S. Swaminathan Research Foundation.

Sainath, P. 2010a, December 8. "17,368 farm suicides in 2009." *The Hindu*. Available at: http://www.thehindu.com/opinion/columns/sainath/17368-farm-suicides-in-2009/article995824.ece

Sainath, P. 2010b., December 27. "Of luxury Cars and lowly tractors." *The Hindu*. Available at: http://www.thehindu.com/opinion/columns/sainath/of-luxury-cars-and-lowly-tractors/article995828.ece

Sainath, P. 2012, July 3. "Farm suicides rise in Maharashtra: State still leads the list." *The Hindu*. Available at: http://www.thehindu.com/opinion/columns/sainath/farm-suicides-rise-in-maharashtra-state-still-leads-the-list/article3595351.ece

Shah, M., R. Rao, and P.S.V. Shankar. 2007. "Rural credit in 20th century India: Overview of history and perspectives." *Economic & Political Weekly* 42(15): 1351–65.

Shah, T. 2007. "The Groundwater Economy of South Asia: an Assessment of Size, Significance and Socio-ecological Impacts." Pp. 7–36 in *The Agricultural Groundwater Revolution: Opportunities and Threats to Development*, edited by Mark Giordano and Karen Villholth. Oxford: CAB International.

Shah, T. 2009. "Climate change and groundwater: India's opportunities for mitigation and adaptation." *Environmental Research Letters* 4(3): 1–14.

Shankari, U. 1991. "Major problems in minor irrigation: Social change and tank irrigation in Chittoor district of Andhra Pradesh." *Contributions to Indian Sociology* 25(1): 85–113.

Shetty, S. L. 2009. "Agricultural Credit and Indebtedness: Ground Realities and Policy Perspectives." Pp. 61–86 in *Agrarian Crisis in India*, edited by D. Narasimha Reddy and Srijit Mishra. New Delhi: Oxford University Press.

Singh, D. K., and A. K. Singh. 2002. "Groundwater situation in India: Problems and perspective." *International Journal of Water Resources Development* 18(4): 563–80.

Singharoy, D. 2004. *Peasants' Movements in Post-Colonial India*. New Delhi: Sage.

Sreedhar, G. 2007. "Tank Irrigation Management at the Peril: Emerging Issues and Challenges." Pp. 228–44 in *Irrigation Development: Issues and Challenges*, edited by P Narashima Rao. New Delhi: Discovery.

Sreedhar, G. 2010. "The Drought Prone Anantapur District of Andhra Pradesh." Pp. 219–41 in *Agrarian Crisis and Farmer Suicides*, edited by R.S Deshpande and Saroj Arora. New Delhi: Sage.

Swain, A., and O. Charnoz. 2012. "In Pursuit of Energy Efficiency in India's Agriculture: Fighting 'Free Power' or Working with it?" Pp. 1–80 in *Working Paper 126*. Paris: Agence Française de Développement.

Swyngedouw, E. 2004. *Social Power and the Urbanization of Water: Flows of Power*. Oxford: Oxford University Press.

Taylor, M. 2011. "'Freedom from poverty is not for free': Rural development and the microfinance crisis in Andhra Pradesh, India." *Journal of Agrarian Change* 11(4): 484–504.

Taylor, M. 2012. "The antinomies of 'financial inclusion': Debt, distress and the workings of Indian microfinance." *Journal of Agrarian Change* 12(4): 601–10.

Vakulabharanam, V., and S. Motiram. 2011. "Political Economy of Agrarian Distress in India since the 1990s." Pp. 101–26 in *Understanding India's New Political Economy: A Great Transformation?*, edited by Sanjay Ruparelia, Sanjay Reddy, John Harriss, and Stuart Corbridge. London: Routledge.

van der Ploeg, J. D. 2008. *The New Peasantries: Struggles for Autonomy and Sustainability in an Era of empire and Globalization*. London: Earthscan.

Vasavi, A. R. 2012. *Shadow Space: Suicides and the Predicament of Rural India*. New Delhi: Three Essays Collective.

World Bank. 2008. *Climate Change Impacts in Drought and Flood Affected Areas: Case Studies in India*. Washington, DC: IDRC.

World Bank. 2010. *Deep Wells and Prudence: Towards Pragmatic Action for Addressing Groundwater Overexploitation in India*. Washington, DC: IBRD.

8 Mongolia – pastoralists, resilience and the empowerment of climate

In the summer of 2013, the Asian Development Bank (ADB) produced a short video to illustrate the impacts of climate change on Mongolian pastoral herders. The immediate motivation for this film was to showcase the bank's programmes designed to foster resilience among rural communities that had suffered catastrophic livestock losses in a series of winter storms (*dzuds*) occurring in 2009–2010.[1] Against the striking visual imagery of horseback riders herding cattle through vast expanses of grassland, the video opens with a commentary about how the climate change impacts threaten the nomadic lifestyles of pastoralists that are largely unchanged from the age of Genghis Khan. The symbolism of the video is clear. In a land of harsh extremes, from blistering summer heat to unrelenting winter cold, climate is an implacably powerful force upon the Mongolian steppe. Under the spectre of global climate change, the projection of increased extreme weather events presents a new and daunting threat to pastoral livelihoods and their associated forms of natural resource management. Such herders, it is projected, are uniquely and deeply vulnerable to climate change.

For the ADB, a series of key solutions springs seamlessly from this framing. With the central problem identified as a tension between herder livelihoods and an increasingly capricious climate, building resilience centres upon a series of technical fixes and institutional reforms at the level of herder practices. These involve teaching herders to grow more drought-resistant fodder varieties and to circulate their animals more rapidly between pastures, incorporating them into livestock insurance schemes and – most fundamentally – creating programmes for livelihood diversification that move rural communities away from a dependence on livestock (Asian Development Bank 2014). This emphasis on building localised resilience stretches beyond the ADB. A more comprehensive and nuanced set of recommendations is provided by a World Bank-commissioned report, one that is based in part from a series of localised studies and interviews with herders (Fernandez-Gimenez, Batjav and Baival 2012). In this document, we find an itinerary for a systematic series of institutional reforms that are projected to provide a better framework for herders to collectively manage common property and facilitate communal learning as a means of adaptation. While the emphasis is placed less on technical fixes, adapting to climate change remains similarly represented in terms of pastoralists changing their localised herding practices, developing better coordinating institutions and diversifying their livelihood portfolios (see also, Batima et al. 2008). The overall goal is to build more resilient communities that can better weather the impacts of climatic change.

There is, of course, no doubt that climatic change will have important ramifications for herder livelihoods and that herder practices will necessarily undergo transformations. To address the socio-ecology of nomadic pastoralism at this point of friction, however, is to marginalise the complex historical forces that have produced this moment of tension on the steppe. I argue that, notwithstanding the representation of Mongolian pastoralism as a persistent and traditional form of livelihood – a depiction that forms part of a common yet entirely modern narrative by the Mongolian state (Marin 2008; Sneath 2010) – it is extremely problematic to understand pastoral livelihoods in this way. Pastoral livelihoods are inherently situated within and shaped by a wider set of socio-ecological dynamics and institutional transformations. In contemporary Mongolia, the experience of climatic change by pastoralists is realised through the specific historical experience of de-collectivisation since the 1990s and the contemporary shift towards a market-orientated and extraction-based economy. Notwithstanding its localistic and presentist approach to climate change adaptation sketched above, the ADB understands this point well. Like the World Bank, it has played a key role in promoting these dramatic transformations of Mongolian society over the past quarter century and is aware of their major implications for how pastoral livelihoods are enacted (Sneath 2003).

In what follows, I historicise Mongolian pastoralism and climatic change by charting how, over the past two decades, deep institutional reforms transformed the numbers of pastoralists, the size and composition of their herds and the ways in which they orientate their herding practices. By promoting the individualisation and marketisation of pastoralism, macro-policy trends have greatly expanded the social differentiation of herder households, the expansion of herd sizes and particularly a dramatic increase in goat herding aimed at foreign cashmere markets. Cumulatively, such transformations have created a very different socio-ecology of pastoralism on the steppe, in which climatic extremes play a more potent role. I therefore seek to invert the question of adaptation. Instead of focusing on how to make herders more resilient to climate, we might first ask why climate has been so empowered as an agent of social change. On this basis, I argue that the issues facing contemporary Mongolian pastoralism cannot be reduced to the relationship between pastoralists and an omnipotent climate. To do so, I maintain, is to fundamentally de-historicise – and simultaneously de-politicise – the complex forces shaping herder livelihoods.

Climate, *dzud* and the steppe

The current anxiety regarding the future of Mongolian pastoralism within the context of climate change was brought to the fore by the tragic events of the winter of 2009–10 wherein Mongolia experienced a period of extreme winter weather known as a *dzud*. Heavy snowfall, low temperatures and strong winds over much of the country created a situation in which more than 8.5 million livestock – approximately 20 per cent of the country's livestock – perished. This disaster directly affected most of the herding community, comprising 769,000 herders and their immediate families, or around 28 per cent of Mongolia's population. Accounts suggest that some 44,000 households lost all of their livestock and 164,000 lost more than half their herd, making it the worst *dzud* in terms of its impact on animal mortality and herder livelihoods since 1945 (UNDP 2010; Fernandez-Gimenez, Batkhishig and Batbuyan 2012).

Although a *dzud* is sometimes referred to as a climatic event or a natural disaster, its determinants are emphatically socio-ecological. The title of *dzud* covers a range of meteorological phenomena that interact with herding practices over the course of a year. In this respect, the *dzud* of 2009–10 was a 'white *dzud*' in which heavy snowfall over a wide area blanketed the ground with an impenetrable barrier that prevented livestock from accessing grass or other grazing matter.[2] The resulting inability of animals to feed, along with exposure to cold and wind, took a heavy toll on herds, leading to high rates of mortality. While *dzuds* manifest themselves as a winter phenomena, it is the occurrence of such adverse conditions immediately following summer drought that is a core factor in animal mortality. When drought occurs upon the steppe within summer months there is a tendency for soil moisture content to decline and grass to become patchy. This lowers the biomass productivity of the grasslands, reduces summer grazing potential and restricts livestock weight gain, leaving animals more vulnerable to extreme winter conditions.[3] The period of intense summer droughts between 1999 and 2002, for example, directly contributed to particularly severe *dzud* events in the winters of these years (Batima et al. 2005) and this occurred again in 2009.

Owing to its projected influence upon both the summer and winter meteorological dimensions of *dzuds*, climate change is identified as a serious threat to pastoralist livelihoods. Over the past half century, Mongolia has seen a shift in a number of climate registers, with the average yearly temperature increasing by at least 1.80 C since 1940 and adverse meteorological events – including winter storms, drought periods and extreme temperatures – becoming more frequent (Dagvadorj et al. 2009). These meteorological trends are seen to impact directly on the production of biomass on the steppe owing to an increase in grassland aridity. Less rain – or, more pertinently, precipitation that occurs in concentrated bursts that cannot be absorbed by the soil – is anticipated to result in less dense pasture coverage. At the same time, rising temperatures are projected to impact localised hydrological cycles through greater transpiration, the melting of high mountain glaciers and the intensive degradation of permafrost. For some regions, such trends are anticipated to result in a likely decline in soil moisture with a corresponding impact on plant growth over the spring and summer seasons. Over the next half century, the boundary of the Gobi desert is expected to move north at an average speed of 6–7 km a year, a trend that will potentially have limiting effects on the availability of pasture resources in the future (Bayasgalan et al. 2009: 21).

Notwithstanding some important regional variations, there is no doubt that meteorological shifts will have a direct impact on the socio-ecology of livestock herding on the steppe.[4] Pastoralists, however, do not simply circulate across a given 'natural' environment that changes beneath their feet owing to meteorological drivers. Rather, they play a key role in producing lived environments through their collective herding practices. The number and kinds of animals they stock in their herds and how they move them across the land according to specific seasonal rhythms greatly shapes the biotic qualities of the landscape (Tsagaan Sankey et al. 2012). Concentrated grazing affects the soil, vegetation, water resources, insects and other animal species that form part of the steppe. Prolonged heavy grazing in one location, for example, changes the types of plant species that predominate in an area and shapes the number and diversity of pollinating insects, which can cumulatively create enduring changes in the land (Yoshihara et al. 2008). Grasses and shrubs frequently decline in number while other plant species, such as variants of Artemisia that livestock find distasteful, come to predominate (Fujita and Amartuvshin 2013). Moreover, the specific animal composition of herds also has determinant effects on the landscape. Camels and cattle tend to consume high-growing vegetation, whereas sheep and particularly goats are active foragers and graze more intensively from top to the bottom of plants. A heavy concentration of goats in herds, for example, leads to a more intensive form of grazing that often destroys plants down to the root (Berger, Buuveibaatar and Mishra 2013). Soil degradation, erosion and alkalinisation can all emerge from concentrated grazing practices, and this in turn can aggravate a drop in soil moisture content and promote the onset of desertification.

The biophysical character of the steppe environment, therefore, is actively produced by human activities. Humans circulate concentrated grazing herds through the steppe that engage in complex ways with forms of plants, other animal species and sources of water. Pastoralists, therefore, are significant agents of environmental change and the grasslands of the Mongolian steppe are, in part, anthropogenic. This central role in the collective production of their lived environments is widely recognised by herders within what is termed their 'traditional' forms of knowledge (Fernandez-Gimenez 2000; see also, Ingold and Kurttila 2000). Precisely how pastoralists engage the land and shape the environment, however, is deeply interwoven into the broader political ecology of which they are part (cf. Dove 2004). Pastoral livelihoods have historically been shaped by diverse and multi-scalar flows of people, animal products and finance between rural and urban areas. Such networks encompass both regional and national dimensions but are also linked into flows of goods and finance that stretch much further afield (Honeychurch 2010). From the fluctuating international cashmere market to the concerted attempts to mass-extract metals and minerals from under the steppe's soil, Mongolian herders find themselves part of productive processes that are global in scale and have manifold and complex impacts upon the lived environments of the steppe.

In this respect, we need to be profoundly cautious when approaching deterministic narratives of climatic change in contemporary Mongolia. While Batima and co-authors (2008) estimate that under climate change projections the live weight of summer-born cattle and ewes is likely to decline strongly, therein opening the grounds for adverse winter weather to extract a heavy toll on livestock; we might first question how the steppe has been produced in a way that makes climate act in such an apparently deterministic fashion. This forces us to address a much wider scope of socio-ecological change. In 1990, for example, the goat population in Mongolia was around five million. By 2009, immediately prior to the onset of *dzud*, it was almost 20 million, a four-fold increase. The overall population of livestock animals, moreover, had risen from around 25 million to 44 million over the same period, creating a massive shift in the socio-ecology of the steppe. This raises pivotal questions. What social forces had driven herd numbers – goats in particular – to increase so dramatically and how did this reshape the lived environment? How did herders find the means to expand livestock numbers and how did the *dzud* impact differently on distinct strata of the herding population depending on their insertion into broader social division of labour within Mongolia? The relationship that herders have to changing meteorological forces, therefore, is one steeped in history.

From empire to collectivisation

At a time that many are questioning the immediate future of Mongolian pastoralism, archaeologists are pointing to its redoubtable character over the longue-durée. As William Honeychurch notes, Mongolian herders have been at the nexus of roughly 2,000 years of long-distance exchange, transport logistics, communications and inter-cultural dialogue founded on mobile pastoral production (Honeychurch 2010).

From the sixteenth century until the twentieth century, herders were incorporated within a quasi-feudal context in which hereditary lords governed administrative districts as part of a broader Chinese empire. These authorities organised land use within their districts, in which they facilitated a sophisticated series of herder movements across selected pastures in a way that targeted specific ecological niches through the season. The semi-feudal context, however, meant that herding was tied into a structured series of social relations through which services and goods were steadily rendered to the nobility and religious establishment (Himmelsbach 2012). As a result, while herders often had large numbers of livestock, the bulk of herds these did not belong directly to them but were managed on behalf of the nobility, ecclesiastical authorities or merchants. Herders provided their labour to the nobility and church by taking care of their animals and delivering products at given times. Moreover, by the later nineteenth century, a sharply deepening indebtedness of herders to Chinese merchants led to a new de-facto merchant ownership over a substantial segment of the Mongolian herds. This indebtedness put intense strains on the social foundations of herding practices. A contemporary observer noted that Chinese traders were taking an annual toll of 25,000 horses, 10,000 cows and 250,000 sheep from one single district alone (Sneath 2010).

This form of semi-feudal pastoralism was only broken in the 1920s, when socialist revolution in the context of regional political upheavals tamed the power of merchants, nobility and the church. The livestock owned in absentia by the nobility were generally turned over to the herders that maintained them and the latter fiercely protected their newfound autonomy. While state authorities, operating under Soviet ideological and political influences, attempted to collectivise herding in the late 1920s and 1930s, sustained herder resistance repeatedly frustrated such efforts (Bruun 2006). Notwithstanding its early failure, collectivisation proceeded apace during the 1950s within the context of a more consolidated state socialism. By 1959, the vast majority of herding households were incorporated within larger herding collectives that controlled more than 70 per cent of the national livestock and which were closely knitted into the wider structures of the command economy.

This dramatic reorganisation of herding reshaped the socio-ecology of the steppe. First and foremost, collectivisation was a means to subject pastoralists to a rationalisation of their practices, including their nomadic movements and the composition of herds, that ultimately sought to subordinate them to a strong productivist ethos (Mearns 1996; Upton 2010). Within the socialist system, herding was seen as an integral part of a broader economic transformation that would tie livestock production to a range of urban-based industries. The resulting emphasis on economies of scale and scientific management within livestock production was reflected in the gathering of herds into ever-larger collective units (*negdels*) that were often devoted to a single species of livestock. By the 1980s, most districts contained only a single collective, and the chairman of the herding collective also served as district governor (Sneath 2003). This structure tended to partially undermine the informal institutions that had previously governed pasture usage and nomadic movements. Robin Mearns, for example, notes how the political exigencies of enforcing control over collectives often led to constraints on some seasonal migratory practices (*otor*) that had longstanding basis in pastoral management (Mearns 1996; also Upton 2012a). At the same time, a more localised system of circulating between pastures was institutionalised under the *negdel* system and facilitated through the generalisation of state-owned trucks for transportation. This provided an important means of overcoming localised drought or *dzud* (Fernandez-Gimenez 1999b; Upton 2010).

In terms of ownership, under the collective system herders no longer managed their own animals beyond a small reserve of around 50 that remained tied to households. Instead, they were integrated into the collective management of state herds that incorporated some 70 per cent of national livestock. In return for effectively becoming labourers within the collective, herders received a salary and a range of social provisions. While collectivisation inherently constrained herder autonomy, the *negdels* did provide some notable benefits in terms of the security they provided. The production of meat, wool, milk and other livestock products was integrated into a managed system of guaranteed prices for collectives who then paid a regular salary to members. Concurrently, collectivised forms of seasonal transportation, shelter provision, pump drilling and fodder production and veterinary services provided a coordinated and relatively inclusive means of organising the wider institutional basis for livestock management. At the same time, the Mongolian state developed a strongly embedded set of policies to provide health and education services to nomadic herders and achieved virtually universal literacy (Fratkin and Mearns 2003). While pastoral livelihoods remained materially sparse, absolute poverty was virtually unknown (Mearns 2004b).

The 'era of the market'

The crisis of the Soviet bloc in the late 1980s, however, created grave problems for Mongolia. With the collapse of a key political and financial backer, the country faced a sharp reduction in aid from Moscow and the disruption of its primary trading linkages. Although Western financial institutions stepped into the ensuing fiscal gap, they did so with the aim of not simply stabilising Mongolia but to fundamentally reshape the country's institutional framework. Under the auspices of the World Bank, International Monetary Fund and ADB, structural adjustment programmes were established to break down the state-orientated economy through an extensive programme of liberalisation and privatisation to facilitate a liberalised market economy. Despite the recognition that such programmes would involve short-term pain – a process of 'creative destruction' wherein market forces were expected to redistribute resources in accordance with a perceived economic rationality – they were nonetheless heralded as a singular cure for societies such as Mongolia that demonstrated the strong legacies of a statist political order (see Taylor 2010). On this basis, the post-transition Mongolian state began a standardised reform process of privatising public assets, liberalising prices, cutting state subsidies and expenditures, ensuring currency convertibility and seeking to actively introduce a marketised form of social relations across Mongolian society (Choikhand 2006).

As widely expected, the outcome of these reforms was a dramatic period of socio-economic turmoil under conditions of austerity and state retrenchment. Adjustment led to the generalised collapse of the formerly protected industrial sector, a massive decline in urban employment, a period of high inflation and a significant increase in poverty in both urban and rural spheres (UNDP 2003; Reinert 2004; Nixson and Walters 2006). Various forms of manufacturing and industry that had been constructed during the socialist period disappeared, with the only areas of growth in the decade following the advent of liberalisation being mining, alcohol production and raw materials, including cashmere and bird down (Reinert 2004). At the level of the state, political decentralisation of responsibility for providing services was not matched by fiscal decentralisation (Mearns 2004a). This led to a stark decline in service provision, particularly outside of the capital city, Ulaanbaatar. Mongolia faced a period of dramatic austerity, widening gaps between rich and poor and a sharp rise in systemic corruption as private control of resources became a vehicle for rent seeking among a privileged minority (Sneath 2006).

For herders, the dismantling of the collective institutions of the socialist period created a significant transformation of pastoral livelihoods. The dismantling of the socialist institutions that regulated the herding economy led to a privatisation of the collective herds of sheep, goats, cattle and horses. The latter were partitioned between former members, along with other collective assets, such as motor vehicles, machinery and equipment. At the same time, many urban households that lost state jobs were granted animals as part of the privatisation of state herds. This led to a re-pastoralisation of the economy. Households directly dependent on herding swelled from around 18 per cent of the population in 1988 to as much as 50 per cent merely a decade later (UNDP 2003). For these 'new' herder households, there were strong constraints to developing seasonal movements that could effectively circulate animals among pasturelands. New pastoralists were often shut out of the informal institutions that governed pasture use, campsites and drinking wells and others had little of the accumulated knowledge of sound herding practices that characterised longstanding pastoralists (Fernandez-Gimenez 2000: 1322). For some households the solution was to house their animals with kinfolk in rural areas, a mutually beneficial arrangement that consolidated and expanded a flow of reciprocal goods and services between rural and urban areas (Fernandez-Gimenez 1999a). Others, however, were forced to remain close to the cities and towns, and these localised pastoral practices tended to exacerbate regional grassland degradation owing to limited seasonal mobility.[5]

Alongside vastly increasing the ranks of the herding population, decollectivisation simultaneously removed the formal institutions that had regulated the use of seasonal pastures and provided collective of transport to migrate herds. With the dismemberment of the collectives, individual herders found themselves in a relatively atomised social terrain, loosely bound by informal practices yet operating in a climate of generalised austerity and multiple new entrants to the herding landscape. This caused major disruption for many existing herders who faced new difficulties marketing their produce – formerly undertaken by the collective – and who no longer enjoyed the security of collectively provided hay or transportation (Sneath 2004: 174). The practice of *otor* – the seasonal migration of herds to access pasture – became more tenuous in the context of a lack of transportation and uncertain rights over distant pastures. Some of these informal practices were taken up by *khot ails* – small groups of cooperating households often linked through kinship that acted as a social safety net for poorer households, providing forms of mutual assistance and pooling risk between households, including sharing food resources as well as long-term loans of livestock (Mearns 1996). But such mobility and support also became increasingly defined along class and kin lines. Poor herders faced considerable constraints upon their mobility and they, alongside the new herders that joined pastoralist ranks after the mass-privatisation, were more likely to trespass on the campsites of others and to have their own campsites trespassed upon (Fernandez-Gimenez 1999b).

The question of rights to pasture, therefore, was a central question within post-socialist Mongolia. Despite pressure from international institutions such as the ADB to fully privatise land, successive Mongolian governments maintained land as state owned and – in practice – organised according to informalised common property frameworks.[6] One key element of the 1994 land law, however, was the provision of privatised leases for seasonal campsites that included the grazing lands that surrounded them (Fernandez-Gimenez and Batbuyan 2004; Sneath 2004; Upton 2010). By granting certificates to use pasture surrounding specific campsites, the state effectively created a microcosm of private rights of land usage within a wider system of pastoral movement. This resulted in new forms of exclusion within a generalised framework of common property. Such practices, however, remained entangled in complex informal understandings of usufruct rights in which common rights to pasture stood awkwardly alongside creeping de facto privatisation of campsite areas, leading to tensions among herders.[7] As Caroline Upton notes, the central principal of reciprocity over pasture use in times of need ran in tension with the desire to protect customary winter pastures wherever possible (Upton 2012a). With the expansion of both the number of herders and the size of herds, the institutional vacuum created a tense socio-political situation. While more powerful *khot ails* have proved better able to 'capture' pastures by maintaining camps in several locations at once, marginal herders can face informal yet systematic exclusion (Fratkin and Mearns 2003; Konagaya and Maekawa 2013).

For Mearns, the fractured institutional context governing land usage in this period represented an "abdication of public administration rather than decentralisation or purposive intervention" (2004a: 139). Such neglect of the pastoral sector reflected the concentration of governmental energies on Ulaanbaatar, the capital city, at the expense of rural areas. Indeed, the Mongolian government has been determinedly interested in the development of a 'livestock sector' in contrast to sustaining 'pastoral livelihoods', which manifests an important difference in emphasis and suggests that the process of strong herder differentiation is one that it tacitly supports as part of a putative modernisation process (Marin 2010).

Social differentiation, risk and the growth of herds

The immediate impact of herd privatisation was the creation of a large number of households with small herds, upon which they became directly dependent for their livelihoods. Some 60 per cent of the rural population had herds of below 100 animals and this has consolidated a new dynamic of social differentiation (Batima, Natsagdorf and Batnasan 2008: 80). As a World Bank synthesis report on the livestock sector notes, herders can be understood to fall within four tiers (World Bank 2009). A few very successful herders have accumulated livestock of between 500 to more than 1,000 head. Comprising six per cent of households, these are well-resourced households that generate full-time employment for kin members and often hire casual labourers from poor households. They actively maintain good access to pastures and inputs, build permanent structures upon pastures that they secure for hay and fodder production and possess the transportation to move their herds long distance if and when required. Such large commercialised production units, however, are the exception. A middle range of herding households – comprising about one quarter of pastoralists – have sheep and goats numbering a few hundred complemented with a smaller number of larger animals, such as cattle and horses. At the bottom of the rural hierarchy, poor households typically have anywhere from 100 animals to just a few dozen sheep and goats. The latter frequently labour as hired hands for the richer herding class (World Bank 2009; see also, Sneath 2012: 467).

Households with dramatically different herd sizes play distinct roles within the steppe and have diverging levels of power to shape the conditions under which they operate. This raises a critical point when we consider contemporary discussions of building herder resilience to climate change. The breakup of the collectives involved a massive individualisation and displacement of risk within rural Mongolia, and these risks are sharply divided according to class fractures (Templer, Swift and Payne 1993; Sneath 2010). Two trends stand out. First, as the collective provisions of the *negdel* disappeared, assets and services such as transportation, fodder production and veterinary services fell into private hands and individual acquisition occurred primarily through market means. As a consequence, liberalisation and privatisation not only shifted the terms of livelihood trajectories but simultaneously re-configured the types of social relations and dependencies through which such livelihoods were realised:

> Relations of mutual help became entangled in monetised logics, and since many people now needed cash, the most common requests for help became appeals to borrow money. Almost everyone, then, became enmeshed in monetised webs of obligation now quantified in terms of cash.
>
> (Sneath 2012: 460)

Second, under this individualisation of risk and monetisation of social logics, smaller herders saw livestock as a buffering asset between subsistence and destitution. Without a stable source of income, poorer herders desperately sought to increase their herd sizes as a form of building assets and providing individualised security (Suzuki 2013). Increasing herd sizes, therefore, became an essential livelihood strategy and animals played multiple roles in the post-transition political economy. Not only did they provide a means of meeting direct subsistence needs, they formed a vehicle of commercial investment and facilitated herder engagement with networks of resource exchange that knitted together rural and urban survival strategies.

As a result, the stocking rates of Mongolia's pastoral economy increased dramatically between 1994 and 1999. Whereas the overall number of livestock within Mongolia remained relatively consistent during the collective period, oscillating between 23 and 25 million head from 1970 to 1994, in 1999 it reached a peak of 35 million. When combined with the relative lack of mobility of some sections of the herding population, the development of this concentrated grazing power raised significant concerns about a generalised degradation of steppe pasture (Himmelsbach 2012). As Sternberg notes in his study of pastoralists in the southern Gobi region, livestock numbers tend to display independence from climatic or ecological variables, implying association with "economic motivators rather than environmental factors" (Sternberg, Middleton and Thomas 2009: 375). Notably, a short period of relatively favourable meteorological conditions in the mid-1990s abetted this expansion of herding as a survival strategy. This over-leveraging of the steppe created a hazardous situation in which meteorological forces were greatly empowered. Upton cites an older herder on this shift in which the forces of climate had seemingly taken on a deterministic role:

> . . . in the collective times, there were no serious problems with the pasture . . . (but) now we are just under the influence of nature, we can't do anything against nature . . . now there is dzud in 13 aimags of Mongolia and people are left with no livestock.
>
> (Upton 2010: 867)

This idea of herders being rendered under the influence of nature captures precisely the taught socio-ecological dynamics at play. Through a combination of interlocking historical processes, pastoral livelihoods had been produced in a way that empowered climate to act as a newly decisive force on the steppe.

Enter the *dzud*

The manifest contradictions of this tenuous socio-ecology expressed themselves in a brutal form in 1999–2002, wherein a series of summer droughts followed by winter storms ravaged the expanded number of livestock that populated the grasslands. *Dzuds*, of course, are a common feature of the steppe, occurring on average every three to seven years. Coping with such extreme weather, therefore, is well engrained within pastoral practices, wherein seasonal migration and livestock mobility function as both preventative means to avoid over-grazing and coping strategies in the event of inclement weather. The socio-ecology of post-1992 herding, however, placed significant limits upon such strategies owing to the lack of infrastructure and organisation to facilitate the moving of herds. In this respect, it was smaller herd owners – particularly those new herders that had returned to the land in the initial era of privatisation – who were the worst hit. For herders operating outside the context of *otor* by concentrating their herds on the overgrazed areas close to settlements, drought caused a marked drop in the summer weight of animals. In conditions of harsh winter storms, and with limited abilities to move animals owing to their relative marginalisation from pastures, campsites, transport and fodder production, many poorer households found their entire stock of animals wiped out. As Templer, Swift and Payne (1993) note, *dzud* events disproportionately impact poorer households who tend to lose a greater percentage of their herds. Over this three-year stretch, the number of pastoral households with herds under 100 head of animals declined by about one quarter (World Bank 2009).

The 1999–2002 *dzuds*, therefore, brought the contradictions of the immediate post-transition era to a climax. Successive harsh winters and dry summers triggered cumulative livestock loses of some 8–10 million animals or around 30 per cent of the national herd. At the time this was estimated to be the worst "natural disaster" in Mongolia's recorded history (Batima et al. 2005). The causes, of course, were anything but natural. As Stephen Devereux (2006) noted in his research among Somali pastoralists in Ethiopia, it is not primarily meteorological forces that make pastoralists vulnerable but the increasing marginalisation of their drought-response mechanisms within a broader policy framework. While the collectivised system may have collectively absorbed such losses, under the new system there was an individualised decentralised of responsibility for a socio-ecological disaster whose constitution was entirely beyond the level of any individual herder. The deaths of livestock initiated a distress flight of herders out of rural areas and back towards the urban realm, specifically Ulaanbaatar, where they accumulated in large *gher* (tent) districts that soon grew to comprise half of the city's population. This contradictory process of rural decomposition is captured by the UNDP in its description of the urbanisation of poverty:

> Poverty affects the traditional core of Mongolian society, the herders, who find themselves increasingly marginalized and vulnerable to the vagaries of market forces and of human-induced environmental changes. In response they are migrating to the cities, particularly to the capital, Ulaan Baatar, to escape diminished livelihoods and lack of services. This is gradually shifting poverty to urban centres while weakening further the potential for economic sustainability in rural areas.
>
> (UNDP 2011: xii)

For the Mongolian state, the crisis of pastoralists – and particularly the poorer sections of herding society – represented an unfortunate moment of an otherwise intractable historical process. The *dzud* was used to catalyse a narrative that herding formed no substantive part of Mongolia's future. As the Mongolian Prime Minister Nambaryn Enkhbayar opined in 2001 at the height of rural distress, "In order to survive we have to stop being nomads . . . We have set this agenda for ourselves and I think a lot of herdsmen are starting to understand this is inevitable" (cited in Honeychurch 2010: 409). While it ignored the very modern constitution of the *dzud* as disaster, this portrayal of herding as an anachronism fitted comfortably within an embedded agenda of mineral and hydrocarbon extraction, not least given the pressing weight of national debt to external financial institutions. Despite such expectations, for many pastoral households changing livelihood was not an option. As a result, although many of the new herders created under privatisation were indeed wiped out by the *dzuds*, others remained firmly within the pastoral sector. Attempting to restock their herds – even at the cost of significant indebtedness – seemed a more attractive option than distress migration to the growing shantytowns of Ulaanbaatar. Marginal herder households, therefore, persisted in considerable numbers and according to the Mongolian Ministry of

Environment by the end of 2004 some 230,000 families owned livestock, with just under 74 per cent of those being entirely dependent on income generated from their herding (UNDP 2011).

Indebtedness and the allure of cashmere

The post-*dzud* reassertion of pastoralism opened a second era of the market in Mongolia in which relations of debt became increasingly important drivers of socio-ecological processes on the steppe. As the earlier chapters have noted, credit plays a double edge role within agrarian environments. For rural households, access to credit can appear as an essential tool to push back against existing social and ecological constraints upon livelihoods. At the same time, the resulting debts can amplify the scale of risks faced by borrowers and can entrench forms of surplus extraction and relations of dependency. For herders in Mongolia, the use of credit is well engrained within seasonal cycles. Households often receive significant cash inflows just twice a year – in the spring through wool sales and in the autumn through sales of animal skins and meat – and this can create a pressing need for regular injections of credit to meet recurrent or unexpected cash needs. With the increasingly marketised provision of services and goods – from veterinary requirements to transportation, health and education – such dynamics have been increasingly pronounced (Sneath 2010).

Following the 1999–2002 *dzuds*, however, pastoralists started to experience a new level of indebtedness as credit became the primary means to rebuild herds (Marin 2008). In particular, the potential for rapid financial returns through cashmere production opened a new vector of credit relations in which credit used to buy goats could be leveraged against existing assets on the expectation of rapid future monetary gains. A project headed by the International Fund for Agricultural Development (IFAD) to help pastoralists rebuild herds in northern Mongolia noted that general restocking was largely being undertaken on the basis of a dramatic increase in debt to banks. This growing indebtedness, moreover, was strongly differentiated among herder classes. Herders with large stocks of animals find negotiating credit at preferential rates easier owing to their accumulated assets and better networks. For marginal herders, in contrast, access to credit is both costly and associated with a much more rigid set of repayment requirements. In the post-*dzud* environment, however, it was this latter section that experienced the most pressing need for credit. As the IFAD notes, marginal herders – and particularly female-headed households – were notably more likely to incur significant debts as part of restocking (IFAD 2007).

This escalating need of pastoralists to take on debt was matched by an expansion in predatory lending practices. Over the 2000s, bank-led microfinance began to penetrate deeper into the Mongolian countryside, wherein commercial banks started competed to extend credit to herders at a cost of between 2 per cent and 3 per cent in interest per month (Sneath 2012). Anticipating projected future revenues stemming from an expanding mining boom, capital had begun to flow into the financial system, and Mongolian banks increasingly saw opportunities to extend credit to a general population operating in an increasingly marketised environment. This included lending to herder households across class differentials, many of whom were anxious to access credit as a way to secure their means of livelihood through purchasing livestock (Marin 2008). A study by the European Bank for Reconstruction and Development in the mid-2000s showed that the penetration of credit had advanced strongly across rural Mongolia over the previous years as competition for rural customers had intensified between the main financial institutions, including Khan Bank, XacBank and Mongol Postbank. The study's authors noted that among a sample of rural women selected specifically with the expectation that they were outside the formal financial system, almost one half already had loans and twenty per cent of those had multiple loans from competing institutions (De Haas 2010).

This expansion of credit/debt relations tied into a second pivotal socio-ecological dynamic: namely, a new increase in livestock holdings alongside the proportion of goats within these herds. Goats have formed a longstanding part of pastoralism alongside cattle, sheep, camels and yaks and their proportion increased moderately in the first post-collectivisation decade. It was following the 1999–2002 *dzuds*, however, that goat herding began expanding at a dramatic rate. Rising from around nine million head in 2002, goats expanded to a peak of 20 million in 2008 (Maekawa 2013: 233). With all tiers of herders seeking to exploit new markets for cashmere largely through Chinese commercial networks, this change in herd composition cut across class differentials. It is marginal herders, however, who have become particularly dependent on goat rearing as a high-risk, high-reward means of attempting to secure a livelihood under adverse socio-economic conditions. The proportion of goats in their herds far exceeds that of larger herders who tend to have a more balanced ratio of animals. The latter maintain this balance partly for environmental concerns and partly because meat prices are more stable than cashmere prices, so a diversified herd provides a form of individualised risk-alleviation. In contrast, owing to their socio-economic vulnerability within the post-transition economy and their growing indebtedness following the 1999–2002 *dzuds*, marginal herders tended to pursue high-risk strategies. The allure of significant returns from cashmere offered a potential way out of the reinforcing dynamic of indebtedness and livelihood precarity. Export-orientated cashmere appeared to point towards a more prosperous future.

The switch to a high proportion of goats within herds, however, exposed two key points of tension. First, it drew herders into uneven relationships of exchange with merchant classes and banks. The inability of herders to exercise control over marketing networks and their physical distance from the centres of consumption added to their lack of transportation makes them highly dependent on the middlemen that link producers with Chinese merchants and then onto international markets (Marin 2008). Concurrently, the price of cashmere – a luxury fashion item – is inherently volatile and subject to noted fluctuations. Second, the dramatic increase of goats as a proportion of overall herd composition put intense pressure on the state of many pastures, threatening to accelerate degradation and processes of desertification. As noted above, goats tend to disturb the pasture's regenerative capacities through their tendency to feed intensively on roots and flowers (Berger, Buuveibaatar and Mishra 2013). As a consequence, cashmere production proved to be an individually rational but collectively destructive livelihood strategy. Ai Maekawa (2013) notes that herders concentrating on goat rearing are quite aware of their collectively irrational actions and face moral contradictions over this livelihood choice as they understand the cumulative effects of goat grazing. Notably, goat raising was legally limited in neighbouring Inner Mongolia in 2004 owing to concerns of environmental destruction from goats digging up grass at the roots (Wang and Zhange 2012).

Collectively, the extension of debt and the impact of goats upon pasture established the grounds upon which meteorological forces could once again become intensely destructive of pastoral livelihoods. Two separate forces followed in quick succession. First, the extension of debt on the back of anticipated cashmere sales disintegrated with the rapid fall of cashmere prices following the 2008 global recession. For herders who had staked their debt on the collateral of their herds, the collapse of the price of cashmere threatened destitution, as they needed to deplete their asset base to pay back loans. This created a vicious cycle of vulnerability and over-indebtedness, particularly as banks had systematically undervalued the animals that were being collateralised (Sneath 2010). Subsequently, in the immediate aftermath of this debt disaster, the 2009–10 *dzud* struck. While governmental forces once again tended to label the 2009–10 *dzud* a "natural disaster", this was to dangerously misrepresent the socio-ecological dynamics at play. Others were less mystified. As one World Bank consultant noted, Mongolian elders that he spoke to claimed that the 2010 disaster was "not a *dzud* of nature, but a *dzud* of carelessness and neglect of nature" (Sayed 2010).

The immediate outcomes of the *dzud* made clear the tenuous socio-ecological foundations of pastoral livelihoods in the second era of the market. Massive livestock losses incurred by herders and their disproportionate impact upon marginal households sparked a cascade of debt defaults that demonstrated how over-leveraged pastoralists had become on credit-driven livelihood strategies. In conditions of mass default, effective ownership of herds appeared to be passing to banks in the way that they had once been transferred to Chinese merchants in the late nineteenth century (Fairclough 2009; Sneath 2010). Notably, more than 80 per cent of the recipients of a European Union-sponsored food aid programme to target relief on *dzud*-afflicted households in western Mongolia suggested that they would need to use the monies saved through aid disbursements to immediately pay back debt (ACF International 2011). As discussed in Chapter Six in the context of post-flood aid expenditures in rural Pakistan, reaffirming credit worthiness and forestalling the seizure of livestock by creditors frequently becomes a central concern of marginal groups who faced immiseration and the destruction of their asset base and anticipate that credit will be central to their survival in the coming months and years.

Without doubt, the many projects to build the resilience of herder communities in the face of this *dzud* may well provide extremely useful forms of localised coordination and coping strategies (Batima et al. 2008; Fernandez-Gimenez, Batjav and Baival 2012). The focus on building stronger collective institutions to regulate herder practices has been reflected in the activities of international development agencies who have allocated US$77.5 million to 14 projects since 2006 centred on the creation and formalisation of more than 2,000 herder groups (Upton 2012a: 163). As Hijaba Ykhanbai (2011) notes, however, while collective community organisations may well be a suitable vehicle to better coordinate localised actions, their effectiveness requires a broader institutional change at a national level. Part of the success of the *negdels* in providing security during the collective period was due to their embeddedness within a wider political economy that valued pastoralism. Since the transition, however, the Mongolian state has shown little interest in such a course. The institutional framework of livestock management, trade and commerce strongly encourages an individualisation of herding rather than community- or cooperative-orientated activities. It also directly encourages the expansion of herd sizes as an individually rational, but collectively irrational, form of livelihood and risk management. Considered an anachronistic residue rather than a source of sustainable livelihoods, there is little recognition of the role that herders could play in maintaining a balanced socio-ecology on the steppe in contrast to, for example, the effects of large-scale mineral extraction. Indeed, the Mongolian state has hedged its fortunes on this latter, far more lucrative, form of socio-ecology.

Mongolia and the metabolism of global capitalism

While contemporary Mongolia might not immediately strike one as being at the heart of world capitalism, it is nonetheless emerging as a centrally important frontier of accumulation owing to the extensive mineral and hydrocarbon deposits buried beneath its grasslands and deserts. In addition to the world's largest undeveloped deposits of copper and gold, Mongolia is estimated to have coal reserves sufficient to fuel every power plant in China for the next 50 years (Hook 2011; Suzuki 2013). It is not surprising, therefore, that the world's largest mining and energy conglomerates are in constant yet frequently tense negotiations with the Mongolian government over rights to extract these resources from the ground. While pastoralists have been migrating outwards from Mongolia's rural limits, nomadic financial capital presently shows great interest in making such regions its home. As a *Financial Times* correspondent provocatively described:

> Outside Ulaanbaatar one Friday night, around a bonfire pop hits pulse through the air and vodka flows unrelentingly as people start to dance to the beat of DJ Zola, who spins out of the back of a truck. The party is one of the pitstops on a tour for fund managers and investors from London, New York,

> Moscow and Zurich who have come to see the opportunities on offer in Mongolia, one of the hottest destinations for resources investment today.
> (Hook 2011: 1)

With several years of double-digit rates of GDP growth, there is talk of Mongolia becoming the new Dubai under a rapid transformation fuelled by hydrocarbon, metal and other mineral exports. Yet while sections of Ulaanbaatar have indeed evidenced this influx of wealth through the presence of mining conglomerates, the impact upon Mongolia's wider population remains uncertain. In rural areas, turning the minerals under Mongolian soil into a frontier of capital accumulation has interfaced with the ongoing struggles over land and livelihoods characteristic of the post-collective period. At the heart of the issue is a fundamental contradiction noted by Sheehy and collaborators. As the scale of unregulated extraction-orientated development and infrastructure construction expands, the rangeland of the steppe experiences a cumulative degradation that is increasingly severe and permanent (Sheehy et al. 2010: 30).

Mining in Mongolia is, of course, a longstanding industry and small-scale informal mining expanded greatly over the 1990s as other livelihood options contracted. Those ejected from the urban economy or from herding became so-called 'ninja' miners and dug into the soil around riverbeds, panning it for gold and other metals in a largely unregulated, dangerous and environmentally destructive cottage industry. The scale of current expansion, however, dwarfs all previous growth and is increasingly driven by large multinational conglomerates. Tavan Tolgoi, for example, is the largest coalmine in Mongolia and is estimated to sit upon the biggest single deposit in the world, encompassing an estimated six billion tons of high quality coke-coal. With a water- intensive open-pit production process, the mine presently produces 10,000 tons of coal per day. Through daily convoys totalling 100 trucks, those materials are transported 270 kilometres across arid desert and grasslands via a largely unpaved set of roads to the Chinese border. There the concentrated hydrocarbon is offloaded to power the industrial metabolism at the heart of the workshop of the world.

Mining projects like Tavan Tolgoi and the equally huge Oyu Tolgoi copper and gold complex attract significant attention because of the relatively politicised struggles surrounding the distribution of their colossal rents and the consequential damage to environment and herder lifestyles. The issues raised by these megaprojects, however, are symptomatic of the widespread expansion of smaller scale mining operations across the country. Mineral extraction commonly requires large amounts of water, while the use of heavy chemicals in mining operations can infiltrate watersheds to show up in both rivers and groundwater (Suzuki 2013). The transport of output across grazing areas, moreover, can also extend grassland degradation and disrupt migration routes, which has caused conflicts with herders over land use claims. Such conflicts, however, are played out in an institutional context that heavily supports extraction (Suzuki 2013). The Mineral Law of 1997, for example, was heralded by the international mining community as one of the most liberal mining laws in Asia (Byambajav 2014). It significantly relaxed rules for obtaining an extraction license and staking claims, thereby creating new forms of enclosure and exclusion that conflicted with the institutions that govern herder pasture rights. Lands that had been presumed as common-rights now appeared to be subordinated to a definitive set of rights governing mineral extraction (Upton 2012b). It is in this context that struggles of local civil society groups seeking to protect the environment and livelihoods (or 'the homeland') of the local people from the threats imposed by mining operations have come to be the most sustained grassroots movement in Mongolia during the past decade (Byambajav 2014). As a result of these pressures, a 2009 law sought to curtail mining in forest areas and in river basins owing to its impact on local ecosystems. It therefore stands an important counterpoint to the earlier legislation. Outside of river watersheds, however, its application is much less certain given the inherent institutional ambiguities over herder land rights (Upton 2012b; Suzuki 2013).

While the expansion of mineral exploitation has raised tensions with herders, it is widely presented as a windfall for the country as a whole. In positioning itself as a raw materials frontier of Chinese industrial transformation – China being the destination for 93 per cent of Mongolian exports in 2012 (UNDP 2013) – there are huge projected profits from these projects. Those gains are likely to fall extremely unevenly. With global mining firms using technology-intensive extraction processes, employment creation is not high and direct rents from commodity extraction tend to fall to a select few. In response to concerted political pressures, including unparalleled street protests in 2008, the Mongolian government has promised to reinvest a significant portion of mining rents through a Human Development Fund, designed in part to invest in social projects and mediate the impacts of rapidly rising prices within Mongolia. In 2011, for example, direct-cash transfers to households from this fund equalled about $180 per year, with one-quarter in the form of tuition fee support. However, in the context of stubbornly high poverty rates and the marginality facing both pastoralists and the expanding migrant colonies in the shantytowns of Ulaanbaatar, there are persistent concerns that these transfers are simply a palliative for the marginal classes in a frontier of global capitalism that currently provides few genuine livelihood prospects (Isakova, Plekhanov and Zettelmeyer 2012).

This returns us to the issue of climate change, to which pastoral livelihoods are presented as being particularly vulnerable and in need of adaptation. As noted above, both the ADB and the World Bank have been avid proponents of building herder resilience to climatic change. Simultaneously, they have backed financially and institutionally the resource extraction boom that directly fuels the release of carbon dioxide into the atmosphere from the Chinese coal-fired power plants to which huge exports of coal are directed. Both China and international mining conglomerates feel that current level of extraction from the massive Tavan Tolgoi coal deposits is a trickle that is waiting to turn into a flood. Graeme Hancock, president of the mining company Anglo American in Mongolia, has said that the size of Tavan Tolgoi dwarfs other deposits globally and his company alone envisages the extraction of 40 million tons of coal annually from the site (Moran 2013). In powering Chinese industrial transformations as part of a global metabolism of production and consumption, hydrocarbon extraction ironically feeds the very climatic processes to which herders are subsequently required to 'adapt'. That these two directly contradictory processes can proceed alongside each other, funded by the same institutional bodies in different capacities, is testament to the power of the discursive separations inherent to the discourse of climate change adaptation. Here we find the production of climate radically separated from its impacts, allowing the same institutional forces to perform the seemingly anatomically impossible act of riding two horses simultaneously in opposite directions.[8] This fundamental contradiction is, of course, not in any way unique to Mongolia. The latter is a microcosm of the fractured political ecology we currently produce on a global scale.

Notes

1 The video is available at http://www.adb.org/news/videos/mongolian-herders-adapting-consequences-climate-change (accessed January 2014).

2 A 'black *dzud*', for example, is where there is too little winter snow, leading to a lack of water for livestock. 'Iron *dzud*', in contrast, reflects a rapid period of winter warming lasting two to five days leads to melting snow cover, yet the ensuing water does not infiltrate the still-frozen ground. The pools this melting creates subsequently freeze, forging ice sheets across the frozen ground surface since the ground is still frozen that lock animals out of grazing and severely limiting their ability to attain food (Batima et al. 2005).

3 This drought-*dzud* nexus does not necessarily apply across the country, with Sternberg raising doubts about the correlation in the extremely arid southern Gobi region (Sternberg, Middleton and Thomas 2009).

4 The 'anthropogenic' component of this climate change is inherently difficult to ascertain. Davi et al. note that the relatively wet conditions of the twentieth century are somewhat anomalous to longer-term climatic trends, suggesting that – in eastern Mongolia at least – a period of drying might be a return to established climatic trends (Davi et al. 2012).

5 While these trends encouraged higher grazing pressure around settlements and urban centres – particularly as these are nexus points at which services are available and markets for production exist – they also affected longstanding herder households outside of large settlements. A study by Lkhagvadorj and collaborators, for example, noted how many herders in the western Altai region began to phase out seasonal migration owing to the high costs involved in transportation and the impacts of rising temperatures, which opened up grazing pastures from earlier in the spring season. They began instead to graze their herds – increasingly composed of goats – on a mix of localised pasture and forest land, yet this has created widespread overgrazing alongside deforestation as villagers increasingly engage in the timber trade to meet immediate cash needs (Lkhagvadorj et al. 2013).

6 The ADB, for example, pushed strongly towards a full privatisation of land, arguing that insecure land tenure arrangements discourages investment in land improvement. Raising the notion of a 'tragedy of the commons', they argued that a failure to privatise land created a lack of positive incentives to increase productivity while exposing land to degradation as, without ownership, users would have limited incentive to avoid overuse (Sneath 2004: 164).

7 It might be noted that localised belief systems link use of land to obligations to spiritual entities who control meteorological forces and can react negatively to a lack of reciprocity in land usage (Sneath 2004: 168; also, Marin 2010).

8 On the UN's World Environment Day, June 5th 2013, the Mongolian President Tsakhia Elbegdorj announced plans to turn Mongolia into a wind power hub of Asia. Funding for this vision would come from the Human Development Fund procured through the export of coal and other minerals. http://www.reuters.com/article/2013/06/05/mongolia-mining-idUSL5N0EH11720130605 (accessed April 10, 2014).

References

ACF International. 2011. *Commodity Vouchers to Support Livestock Herder Disaster Preparedness*. Paris: ACF.

Asian Development Bank. 2014. *Making Grasslands Sustainable in Mongolia: Herders' Livelihoods and Climate Change*. Mandaluyong City, Philippines: ADB.

Batima, P., B. Bold, T. Sainkhuu, and M. Bavuu. 2008. "Adapting to Drought, Zud and Climate Change in Mongolia's Rangelands." Pp. 196–210 in *Climate Change and Adaptation*, edited by Neil Leary, James Adejuwan, Vicente Barros, Ian Burton, Jyoti Kulkarni, and Rodel Lasco. London: Earthscan.

Batima, P., L. Natsagdorf, and N. Batnasan. 2008. "Vulnerability of Mongolia's pastoralists to climate extremes and changes." Pp. 67–88 in *Climate Change and Vulnerability*, edited by Neil Leary, Cecilia Conde, Jyoti Kulkarni, Anthony Nyong, and Juan Pulhin. London: Earthscan.

Batima, P., L. Natsagdorj, P. Gombluudev, and B. Erdenetsetseg. 2005. "Observed climate change in Mongolia." *AIACC Working Paper* 1(12): 1–26.

Bayasgalan, B., R. Mijiddorj, D. Gomboluudev, D. Oyunbaatar, M. Bayasgalan, A. Tas, T. Narantuya, and L. Molomjamts. 2009. "Climate Change and Sustainable Livelihood of Rural People in Mongolia." Netherlands Climate Assistance Programme (NCAP). Available at: http://static.weadapt.org/knowledge-base/wikiadapt/images/c/ca/NCAP.MON.CON-01.BookChapter.final.pdf

Berger, J., B. Buuveibaatar, and C. Mishra. 2013. "Globalization of the Cashmere market and the decline of large mammals in Central Asia." *Conservation Biology* 27(4): 679–89.

Bruun, O. 2006. *Precious Steppe: Mongolian Nomadic Pastoralists in Pursuit of the Market*. Oxford: Lexington.

Byambajav, D. 2014. "The river movements' struggle in Mongolia." *Social Movement Studies* 13(4). Available at: http://www.tandfonline.com/doi/abs/10.1080/14742837.2013.877387#.U9jYuqi5JOQ

Choikhand, J. 2006. "Economic and political liberalization in Mongolia." *The Mongolian Journal of International Affairs* 13(1): 119–32.

Dagvadorj, D., L. Ntasagdorj, J. Dorjpurev, and B. Namkhainyam. 2009. *Mongolia: Assessment Report on Climate Change*. Ulaanbaatar: Ministry of Environment, Nature, and Tourism, Mongolia.

Davi, N. K., N. Pederson, C. Leland, B. Nachin, B. Suran, and G. Jacoby. 2012. "Is eastern Mongolia drying? A long-term perspective of a multidecadal trend." *Water Resources Research* (49): 1–8.

De Haas, R. 2010. "Mongolian microfinance: Some first insights from a randomised field experiment." *EBRD blog*. Available at: http://www.ebrdblog.com/wordpress/2010/02/mongolian-microfinance-some-first-insights-from-a-randomised-field-experiment/

Devereux, S. 2006. *IDS Research Report 57: Vulnerable Livelihoods in Somali Region, Ethiopia*. Brighton, UK: IDS.

Dove, M. 2004. "Anthropogenic grasslands in Southeast Asia: Sociology of knowledge and implications for agroforestry." *Agroforestry Systems* 61(3): 423–35.

Fairclough, G. 2009, April 20. "The global downturn lands with a Zud on Mongolia's nomads." *Wall Street Journal*. Available at: http://online.wsj.com/news/articles/SB124017991210632815.

Fernandez-Gimenez, M. 1999a. "Reconsidering the role of absentee herd owners: A view from Mongolia." *Human Ecology* 27(1): 1–27.

Fernandez-Gimenez, M. 1999b. "Sustaining the steppes: A geographical history of pastoral land use in Mongolia." *The Geographical Review* 89(3): 315–41.

Fernandez-Gimenez, M. 2000. "The role of Mongolian nomadic pastoralists' ecological knowledge in rangeland management." *Ecological Applications* 10(5): 1318–26.

Fernandez-Gimenez, M., and B. Batbuyan. 2004. "Law and disorder: Local Implementation of Mongolia's land law." *Development and Change* 35(1): 141–65.

Fernandez-Gimenez, M., B. Batjav, and B. Baival. 2012. *Lessons from the Dzud: Adaptation and Resilience in Mongolian Pastoral Social-Ecological Systems*. Denver: Colorado State University.

Fernandez-Gimenez, M., B. Batkhishig, and B. Batbuyan. 2012. "Cross-boundary and cross-level dynamics increase vulnerability to severe winter disasters (dzud) in Mongolia." *Global Environmental Change* 22(4): 836–51.

Fratkin, E., and R. Mearns. 2003. "Sustainability and pastoral livelihoods: Lessons from East African Maasai and Mongolia." *Human Organization* 62(2): 112–23.

Fujita, N., and N. Amartuvshin. 2013. "Distribution Patterns of Vegetation as a Fundamental Factor in Mongolian Ecosystems." Pp. 44–50 in *The Mongolian Ecosystem Network: Environmental Issues Under Climate and Social Changes*, edited by Norio Yamamura, Noboru Fujita, and Ai Maekawa. Tokyo: Springer.

Himmelsbach, R. 2012. "Collaborative Pasture Management: A Solution for Grassland Degradation in Mongolia?" Pp. 170–94 in *Change in Democratic Mongolia: Social Relations, Health, Mobile Pastoralism and Mining*, edited by Julian Dierkes. Leiden: Brill.

Honeychurch, W. 2010. "Pastoral nomadic voices: a Mongolian archeology for the future." *World Archeology* 42(3): 405–17.

Hook, L. 2011, July 8. "Boomtown Mongolia." Financial Times. Available at: http://www.ft.com/intl/cms/s/2/7785732e-a84a-11e0-9f50-0014

Ingold, T., and T. Kurttila. 2000. "Perceiving the environment in Finnish Lapland." *Body and Society* 6(1): 183–98.

IFAD. (International Fund for Agricultural Development). 2007. *Mongolia: Arhangai Rural Poverty Alleviation Project Completion Evaluation*. Report #1889-MN. Rome: IFAD.

Isakova, A., A. Plekhanov, and J. Zettelmeyer. 2012. "Managing Mongolia's resource boom." *European Bank for Reconstruction and Development Working Papers* 138(1): 1–43.

Konagaya, Y., and A. Maekawa. 2013. "Characteristics and Transformation of the Pastoral System in Mongolia." Pp. 33–43 in *The Mongolian Ecosystem Network: Environmental Issues Under Climate and Social Changes*, edited by Norio Yamamura, Noboru Fujita, and Ai Maekawa. Japan: Springer.

Lkhagvadorj, D., M. Hauck, C. Dulamsuren, and J. Tsogtbaatar. 2013. "Pastoral nomadism in the forest-steppe of the Mongolian Altai under a changing economy and a warming climate." *Journal of Arid Environments* 88(1): 82–89.

Maekawa, A. 2013. "The Cash in Cashmere: Herders' Incentives and Strategies to Increase the Goat Population in Post-Socialist Mongolia." Pp. 233–45 in *The Mongolian Ecosystem Network: Environmental Issues Under Climate and Social Changes*, edited by Norio Yamamura, Noboru Fujita, and Ai Maekawa. Tokyo: Springer.

Marin, A. 2008. "Between cash cows and golden calves: Adaptations of Mongolian pastoralism in the 'age of the market'." *Nomadic Peoples* 12(2): 75–101.

Marin, A. 2010. "Riders under storms: Contributions of nomadic herders' observations to analysing climate change in Mongolia." *Global Environmental Change* 20(1): 162–76.

Mearns, R. 1996. "Community, collective action and common grazing: The case of post-socialist Mongolia." *Journal of Development Studies* 32(3): 297–339.

Mearns, R. 2004a. "Decentralisation, rural livelihoods and pasture-land management in post-socialist Mongolia." *European Journal of Development Research* 16(1): 133–52.

Mearns, R. 2004b. "Sustaining livelihoods on Mongolia's pastoral commons: Insights from a participatory poverty assessment." *Development and Change* 35(1): 107–39.

Moran, T. 2013. "Avoiding the 'resource curse' in Mongolia." *Peterson Institute for International Economics Policy Brief* PB13(18): 1–8.

Nixson, F., and B. Walters. 2006. "Privatization, income distribution and poverty: The Mongolian experience." *World Development* 34(9): 1557–79.

Reinert, E. 2004. "Globalization in the Periphery as a Morgenthau Plan: the Underdevelopment of Mongolia in the 1990s." Pp. 157–215 in *Globalization, Economic Development and Inequality*, edited by Erik Reinert. Cheltenham, UK: Edward Elgar.

Sayed, A. 2010. "Dzud: a slow natural disaster kills livestock – and livelihoods – in Mongolia," in *East Asia & Pacific on the Rise: Making Development Work for All*, a blog edited by World Bank. http://blogs.worldbank.org/eastasiapacific/dzud-a-slow-natural-disaster-kills-livestock-and-livelihoods-in-mongolia accessed January 5, 2014.

Sheehy, D. P., C. M. Sheehy, D. E. Johnson, D. Damiran, and M. Fiemengo. 2010. *Livestock and Wildlife in the Southern Gobi Region, with Special Attention to Wild Ass*. Washington, DC: World Bank.

Sneath, D. 2003. "Land use, the environment and development in post-socialist Mongolia." *Oxford Development Studies* 31(4): 441–61.

Sneath, D. 2004. "Proprietary Regimes and Sociotechnical Systems: Rights over Land in Mongolia's "Age of the Market"." Pp. 161–83 in *Property in Question: Value Transformation in the Global Economy*, edited by Katherine Verdery and Caroline Humphrey. Oxford: Berg.

Sneath, D. 2006. "Transacting and enacting: Corruption, obligation and the use of monies in Mongolia." *Ethnos* 71(1): 89–112.

Sneath, D. 2010. "Political mobilization and the construction of collective identity in Mongolia." *Central Asian Survey* 29(3): 251–67.

Sneath, D. 2012. "The 'age of the market' and the regime of debt: the role of credit in the transformation of pastoral Mongolia." *Social Anthropology* 20(4): 458–73.

Sternberg, T., N. Middleton, and D. Thomas. 2009. "Pressurised pastoralism in South Gobi, Mongolia: What is the role of drought?" *Transactions of the Institute of British Geographers* 34(2): 364–77.

Suzuki, Y. 2013. "Conflict Between Mining Development and Nomadism in Mongolia." Pp. 231–55 in *The Mongolian Ecosystem Network: Environmental Issues Under Climate and Social Changes*, edited by Norio Yamamura, Noboru Fujita, and Ai Maekawa. Tokyo: Springer.

Taylor, M. 2010. "Conscripts of competitiveness: culture, institutions and capital in contemporary development." *Third World Quarterly* 31(4): 561–79.

Templer, G., J. Swift, and P. Payne. 1993. "The changing significant of risk in the Mongolian pastoral economy." *Nomadic Peoples* 33(1): 105–22.

Tsagaan Sankey, T., J. Sankey, K. Weber, and C. Montagne. 2012. "Changes in Pastoral Land Use and their Effects on Rangeland Vegetation Indices." Pp. 151–64 in *Change in Democratic Mongolia: Social Relations, Health, Mobile Pastoralism and Mining*, edited by Julian Dierkes. Leiden: Brill.

UNDP. 2003. *Human Development Report Mongolia*. Ulaanbaatar: UNDP.

UNDP. 2010. *Dzud National Report 2009–2010*. Ulaanbaatar: UNDP/NEMA.

UNDP. 2011. Human Development Report Mongolia, 2011. Ulaanbaatar: UNDP.

UNDP. 2013. "About Mongolia." edited by UNDP. Geneva: UNDP.

Upton, C. 2010. "Living off the land: Nature and nomadism in Mongolia." *Geoforum* 41(4): 865–74.

Upton, C. 2012a. "Managing Mongolia's commons: Land reforms, social contexts, and institutional change." *Society & Natural Resources* 25(2): 156–75.

Upton, C. 2012b. "Mining, Resistance and Pastoral Livelihoods in Contemporary Mongolia." Pp. 223–48 in *Change in Democratic Mongolia: Social Relations, Health, Mobile Pastoralism and Mining*, edited by J. Dierkes. Netherlands: Brill.

Wang, X., and Q. Zhange. 2012. "Climate variability, change of land use and vulnerability in pastoral society: A case from Inner Mongolia." *Nomadic Peoples* 16(1): 68–87.

World Bank. 2009. *Mongolia: Livestock Sector Study Volume 1 Synthesis Report*. Washington, DC: IBRD.

Ykhanbai, H. 2011. *Community Based Co-Management of Natural Resources in Mongolia: Ten Years of Experience*. Ulaanbaatar: IDRC.

Yoshihara, Y., B. Chimeddorj, B. Buuveibaatar, B. Lhagvasuren, and S. Takatsuki. 2008. "Effects of livestock grazing on pollination on a steppe in eastern Mongolia." *Biological Conservation* 141: 2376–86.

Conclusion

Adapting to a world of adaptation

This book has charted how the idea of adaptation has risen to become the conceptual lynchpin of institutional responses to global climate change. Nowhere is this more evident that in the field of international development, where adaptation, with its compelling imagery of impending climatic disturbances and social adjustments, has become the default means through which climate change is conceptualised and ensuing social responses planned. The emphatic call for "adaptation now!" is a recurring thematic in both the policy and academic literature and has elicited a rush to repackage development planning within the conceptual parameters of vulnerability, resilience and adaptive capacity. The rationale for this mainstreaming of adaptation appears to be self-evident. In the context of rising global temperatures and increased incidences of extreme weather, the world's poor – and particularly those in rural areas – are uniquely vulnerable to climatic shifts. As a consequence, adaptation appears to be both an expressly urgent and intrinsically natural response to the very real threats posed by climate change.

Notwithstanding the apparent self-evidence of adaptation, the argument of this book is that we should be extremely cautious of such representations. As Timothy Mitchell once noted, the more natural an object appears, the less obvious its discursive manufacture will be (2002: 210). In this spirit, I have argued that we need to be attentive to the way that the concept of adaptation fashions a closely knit body of ideas around the causes and impacts of climatic change that are seamlessly integrated into prevailing institutional rationalities. Within the adaptation framework, however, there is a strongly engrained tendency to marginalise questions of power and production within climatic change. The chapters of this book, therefore, have collectively sought to de-frame climate change adaptation and the supporting concepts that populate its explanatory framework. In so doing, they highlight a series of core assumptions and contradictions that are built into its analytical framework. While these reside quietly within its conceptual foundations, they fundamentally influence how the paradigm represents the complex relations between meteorological forces and social processes. Two key facets are worth repeating.

First, the seeming naturalness of adaptation stems from the way its framework replicates the essential dualism between society and nature that is deeply embedded in Western social thought. Framed within this Cartesian ontology, the idea of adaptation builds from the notion that climate and society are two distinct domains or systems that engage in a series of external influences. It is this foundational dichotomy that grounds the concept of adaptation and gives it coherence. With climate thrown off-kilter by the impact of anthropogenic emissions of greenhouse gases, climate change is seen to present society with a series of external stimuli, disturbances and threats. Adaptation thereby emerges as a process of planned social adjustments that, by mediating actual and projected climatic threats, can bring the two systems back into a more balanced co-existence.

This imagery of separate systems locked into an endless dance of adaptation, however, seems curiously unsuited to addressing the fundamental questions facing in a world in which humans are active and expansive protagonists in the production of their environments. Far from encountering climate as an external system, we are intimately involved in its very production, just as diverse meteorological forces are unevenly interred within the production of social life. By pre-packaging 'climate' and 'society' as bounded systems, however, the adaptation framework gives no room to explore the ways in which meteorological forces and social energies are intimately intertwined and consistently co-productive. This makes the adaptation framework conspicuously unsuited to the very processes it seeks to explain. As I demonstrated through case studies across a variety of agrarian settings, the idea of an 'external' climate colliding with an 'internal' society becomes impossible to maintain when one considers the historically interlaced processes through which meteorological forces are worked into hierarchically ordered forms of social organisation and physical infrastructures in the ongoing production of lived environments. Far from a clear boundary line between society and climate, in which the former could adapt to the latter, we are faced with complex assemblages of social energies, biological processes and meteorological forces that together produce the world around us. The presence of such uncertain boundaries strongly challenges the very foundations of the adaptation concept because it interrupts the a priori separation of 'internal society' and 'external environment' that animates the paradigm. As the ecological anthropologist Tim Ingold once noted, something must be deeply wrong if the only way to understand our own creative involvement in the world is by taking ourselves out of it (2000: 173).

The second, yet intrinsically related, reason for the salience of climate change adaptation is the way that it accords with existing institutional logics that seek to make climate change governable. The internal–external binary between climate and society noted above may be fundamentally unsuited to the complexities of a world in constant and dynamic transformation, but it is remarkably successful at rendering climate change in a manner legible to governmental forces. It does so by discursively producing of a 'world of adaptation' in which all aspects of society – from households to economies – are portrayed as necessarily adapting to external climatic stimuli. This leads to a framing of climate change in which the latter is represented as an external force that radiates downwards upon a series of seemingly pre-constituted units – societies, communities, ecosystems, economic sectors – that face an exogenous stress to which they must adapt. Adaptive capacity, or the relative ability of social units to respond to climatic stresses, can then emerge as a new and universal object of development upon which the transformative practices of states, institutions and organisations can be set to work. Adaptation, therefore, orders the world in terms of a taxonomy of vulnerability to external climatic threats in ways that strategically elide the central question of how people are rendered vulnerable within the active production of their lived environments and the diverse socio-ecological relationships that underscore such production.

In short, the framework of climate change adaptation abstracts and exteriorises climate from the multiplicity of ongoing transformations that actively remake our world. This act of separation is precisely the means by which climate change can be rendered governable. Such biopolitical intent, however, marginalises essential questions about the embedded forms of power through which humans actively yet unequally transform their lived environments. Indeed, the discourse of climate change adaptation practically mandates us to check such questions in at the door as part of the price of entry. To do so, as Michael Watts (1983) cogently argued in earlier political ecology debates around vulnerability and environmental change, is to create precisely characteristically 'thin' conceptualisations of social-ecological dynamics that risk being both banal and markedly unsuited to addressing the issues at hand.

Producing ourselves differently

Through its examination of socio-ecological change in agrarian environments, the alternative framework put forward in this book sought to open a very different set of questions. First, we need to relinquish the problematic that motivates the politics of adaptation – i.e. how does climate change make rural regions and their populations vulnerable. Instead, we need to put forward a more pointed question: how do we make climate so powerful? This is an important inversion because it puts production back at the heart of the discussion where it belongs. On the one hand, this conceptual shift forces us to examine closely how meteorological forces are actively combined with forms of social organisation, technical infrastructures and biological processes in the ongoing, yet conspicuously uneven, production of agrarian environments. On the other, it raises the question of a stratified global division of consumption in which the increasingly intensive commodification of the human and non-human world has driven forward an expansive industrial metabolism geared towards the relentless accumulation of capital yet hinged to the parallel production of climate at a global scale.

In both cases, far from a clear boundary line between society and climate in which the former could adapt to the latter, the book emphasised the complex assemblages of social, biological and meteorological processes that together produce fragile lived environments in which the relative security of some is directly related to the vulnerability of others. Within the context of agrarian environments, this framework provided a more coherent basis from which it was possible to conceptualise how such socio-ecological forces are both radically productive – fashioning the basic foundations for human life – but also can reach moments of extreme tension that unleash considerable (yet uneven) social destruction in the form of floods, droughts, storms and other socio-climatic events. This focus on the tensions of production is key because it enables us to see that engaging contemporary climatic change is not about adapting to a changing external environment. It is about challenging how we produce ourselves as part of a broader field of socio-ecological processes. As such, it is fundamentally a question of values, politics and power.

In seeking to understand how we make climate powerful, we are therefore able to historicise climate change and write our own creative yet alienated role in its co-production back into our narratives of social transformation. In particular, the focus on humans as active agents of (self)production helps to overcome the prevalent tendency to blame the pressing issues raised by contemporary climate change on the outside threat of an external system unintentionally tipped out of control (Swyngedouw 2010). In contrast, it emphasises the pressing need to collectively deleverage a global capitalist order predicated upon the unending accumulation of productive forces and consumptive practices that give rise to the deadly metabolisms inherent to climatic change and the stark inequalities on which climate is rendered a powerful representative of human insecurity. In a world in which we 'over-produce' nature, and do so in such a brutally unequal fashion, we need to project how we can "slow down fast" simultaneous to reordering existing productive forces towards more equitable ends.

This, of course, raises multiple and challenging questions. As any lecturer in the field of development studies who offers a critical perspective on the contemporary status quo has likely experienced, you are almost certain to be pressed on what 'the solutions' are. Moreover, when providing an incipient range of possibilities for social change, a common refrain is that they are 'not realistic'. By this, the questioner often means that such suggestions are not politically feasible within the current distribution of power across the global political economy. The limit of our political imagination, therefore, crashes headfirst into the barricade of a projected realism.

There is, however, an uncomfortable correlate to this narrative: the lived present seems determinedly unrealistic as a template for the future. Is it 'realistic', for example, to assume that humanity can turn its global population of seven billion into ideal type middle-class consumers, a vision that still animates the telos of development some five decades after Walt Rostow (1960) described it as the final stage of modernisation? The concept of 'ecological overshoot' crudely brings this point home in the way it notes that humanity's annual demand on the natural world has already exceeded what the Earth can renew by more than 50 per cent (OECD 2008). If we accept the paradigm of realism, then we must surely understand that the way we collectively produce ourselves at present is deeply 'unrealistic' given humanity's dependence upon a complex web of socio-ecological processes to which its futures are uneasily tethered. In this respect, I believe that it is important to open up the fundamental premise of development and its teleology of a globalising, boundless consumption (Nigam 2011).

We do not, of course, require the spectre of climate change or Malthussian narratives of finite ecological constraints to challenge the rubric of business-as-usual. We simply need to consider the issue of food production, wherein the globe already produces (and wastes) amply more food than it needs, and yet one in eight people suffer from malnutrition while many more suffer the consequences of a diet predicated upon nutritionally deficient industrial food (Patel 2007; Weis 2007; Winson 2013). This question of agriculture and food cuts to the heart of the many contradictions thrown into stark relief within the context of climate change. As Philip McMichael (2013) notes, the 'agrarian question of food' is a means of thinking about the possibilities for a sustainable socio-ecological future in an era in which climatic change adds further tensions to an already taught political ecology.

The currently institutionalised 'solution' to this issue, as discussed in Chapter Five, is to promote the heightened intensification of agriculture through the generalisation and intensification of industrial farming across the Earth's surface. Unleashing the forces of competition upon agrarian environments, it is estimated, is a necessary process to further the productivity of agriculture by imposing a drive towards efficiency upon a still vast smallholder sector that has insufficiently adopted modern technologies and practices. That such competitive forces undermine the social conditions for a considerable portion of smallholder agriculturists – as seen in all three case studies – is considered an unfortunate but necessary historical task given the latter's economically 'inefficient' use of labour (World Bank 2007). That smallholder farming in an enabling socio-political context may offer a considerably more socio-ecologically efficient manner of producing food and livelihoods while reducing greenhouse gas emissions and promoting ecologically sound farming systems is sacrificed on the altar of technologically driven yield intensification (Altieri and Koohafkan 2008). As Frans Schumacher (1973) once noted, modern economics does not distinguish between renewable and non-renewable materials because its very method is to equalise and quantify everything by means of a money price.

The modernisation project, therefore, is predicated on a model of agricultural efficiency that rests upon the systematic externalisation of costs as a means to maintain productivity (Gliessman 2007; Weis 2007; Altieri and Roge 2011; Martínez-Alier 2011). As the United Nations Environment Program has detailed, industrial agriculture externalises costs to boost yields in two ways. First, it undercuts the ecological underpinnings needed to produce food – water, soil formation, biodiversity – through excessive use, inappropriate management and over-extraction. Second, it is predicated upon the production of side effects – from groundwater contamination, pollution of surface water and greenhouse gas emissions – that undermine the capacity to generate future yields (UNEP 2012, 2014). Indeed, this model of agriculture deepens the very problems that it argues it is responding too. It is fossil-fuel intensive, produces significant methane emissions from industrial livestock production and nitrous oxide emissions from fertiliser application and promotes major land-use changes through deforestation (Martínez-Alier 2011). In short, modern industrial agriculture appears efficient only if we wilfully ignore the externalisation of costs that are displaced outwards across space and time.

Despite the illusory qualities of the current petro-farming, there is no need to presume that such trends will necessarily create an ultimate rupture or systemic collapse. Capitalism has furnished a socio-ecological environment that is deeply conducive to the systematic externalisation of costs, which are then securitised when necessary through the cold logic of coercive force (Taylor 2009). In this respect, industrial agriculture is predicated upon a continual displacement of its core contradictions in a way that paradoxically creates a 'sustainability of unsustainability' (Buttel 1997: 250). The 'externalities' that accompany market-driven production and consumption, therefore, are better represented as a successfully transferred costs for which there exists a strongly girded institutional infrastructure (Martínez-Alier 2002). Outside of strong counter-movements, the resulting symptoms of land degradation, unproductive landscapes and displaced peoples that continue to surface will unerringly be diagnosed as signs of the necessity to deepen the capitalist modernisation project predicated upon market extension, securing property rights and building resilience. This much appeared clear from each of the case studies represented in Chapters Six through Eight. Notably, these trends also are evident in the contemporary 'land grabs' in which control over large tracts of arable land and associated water resources are turned over to corporate management to undertake industrial agriculture on the behalf of foreign consumers rather than investing in sustainable smallholder agriculture (Borras et al. 2011).

We, therefore, need to contest how the future of agrarian environments and their populations is being determined upon a spurious criteria of efficiency determined by a valuation of energy and labour that is entirely contingent and that is unlikely to persist into the near future (Lin et al. 2011). There is, of course, an increasingly voiced alternative to the industrial model of farming and its attendant social-ecology of a streamlined smallholder sector geared towards capital-intensive productivity gains. This is the model of agro-ecology that projects a range of changes in agricultural practices that can underwrite a profound transformation of how we produce, distribute and consume food (De Schutter 2010). If we consider, for example, the types of transformation that the United Nations Commission on Trade and Development special report on agriculture determine as necessary elements of a sustainable future, these range from increasing the soil carbon content through closed nutrient cycles; reduction of greenhouse gas emissions created through industrial livestock production; a stop to current land use change; optimisation of organic and inorganic fertiliser use; a major reduction of waste in food chains; changing dietary patterns; and the reform of the international trade regime for food and agricultural products (UNCTAD 2013). This is not about climate change adaptation. It is about producing ourselves differently.

The vision of agro-ecology, however, is evidently not something that can be simply undertaken through technical changes at the level of the farm. To borrow the terminology of institutional economics, the 'incentive structure' of global capitalism is pervasively misaligned for such a project. As Haroon Akram-Lodhi argues, the minimal conditions for a revitalisation of smallholder agriculture within a global food system orientated towards fair access must include three aspects. First, the redistribution of land and other productive assets outside the immediate logic of 'economic efficiency' in a specifically pro-poor manner across agrarian environments; second, the creation of an entirely different supra-state apparatus that can regulate the movement of food in an anti-corporate fashion that dampens the power of market imperatives by balancing a range of efficiency concerns with investment and employment conditions; and, third, a cultural revolution in food that dismantles a systematically irrational food system directed towards producing nutritionally deficient industrial foods and environmentally destructive farming practices (Akram-Lodhi 2013).

It is, of course, possible to debate the details of Akram-Lodhi's schematic, to focus on the tensions inherent to such a vision and question how such a project might be realised (e.g. Agarwal 2014; Bernstein 2014). Yet, as a set of proposals for collective self-transformation, they help us recognise that we need to re-imagine redistribution as a central pillar of equitable socio-ecological transformation. Redistribution, as Holt-Gimenez and Shattuck note, is not simply about the redistribution of wealth or goods but fundamentally concerns the restructuring of entitlements (Holt-Gimenez and Shattuck 2011: 128). Within agrarian environments redistributive strategies – from land and water rights through to credit policies and subsidies – historically have been a central aim of many agrarian social movements. This is because they allow marginal groups to better position themselves within the active production of their lived environments in a way that improves their livelihood security. Despite their complete marginalisation within the framework of climate change adaptation, these struggles become ever more important within the context of contemporary climatic change. They are not about 'handouts' but asserting democratic control over productive processes that stem the vulnerabilities inherent to agricultural production that are accentuated within capitalist socio-ecologies. There is no circumventing the uncomfortable fact for policymakers that achieving the goals of 'sustainable adaptation' would require not simply a change of relations between marginal groups and the state through ameliorative social policies but simultaneously a form of social transformation predicated upon a fundamental redistribution of power. Thinking beyond adaptation for rural social movements, state bodies and external solidarity groups will be central to turning such possibilities into practice.

On this basis, it is worth noting how the inherent and widely recognised inequities of climatic change potentially open a pathway towards revitalising the idea of redistribution across spatial scales. Such an initiative is manifested in the notion of the 'ecological debt', a concept that seeks to account for how processes of development spatially concentrated in the West have been facilitated and subsidised by the appropriation and degradation of global environmental resources, as well as disproportionate utilisation of the global commons (Foster 1999; Martínez-Alier 2002; Rice 2009). In its discussion of climate change, the World Bank hinted at the competitive appropriation of metabolic processes that underlie the concentrated power to consume, although it refused to develop the point. Switching from SUVs to fuel-efficient passenger cars in the US alone, it notes, would nearly offset the emissions generated in providing electricity to 1.6 billion more people (World Bank 2009: 3). This small point exemplifies the substantive irrationalities operating at the heart of a global division of consumption in which the power of money acts as a claim upon not just material goods for consumption but also the complex metabolic processes that facilitate them in the past, present and future. As James Rice elaborates:

> Ecological flows, when systematically undervalued in monetary terms, displace the environmental costs of this uneven consumption to LCDs (least developed countries). In addition, it allows core countries to inequitably appropriate limited global environmental space or carrying and sink capacity of ecological systems well beyond their own borders. (Rice 2007: 44)

We therefore need to think of redistribution in terms of broad categories of control over productive assets, material goods and socio-metabolic processes. The question of redistribution even in its most basic forms, however, has been forcibly pushed from the active political lexicon in the era of neoliberalism. As Stefan Andreasson argues, "even the suggestion that redistributing existing resources rather than increasing aggregate levels of growth and consumption could improve the lot of the world's destitute is, for most development thinkers and policymakers, not 'viable' (i.e. politically feasible)" (2005: 57). For some, even a modest redistribution of global resources without further expanding production could make major progress in tempering the conditions of insecurity and impoverishment faced by a global majority (Woodward and Sims 2006). Tacitly, of course, there remains an ongoing redistribution of wealth upwards that is inherent to capitalism and is epitomised in the growing affluence centralised in the hands of a global 'one per cent' that has seen real income rise by more than 60 per cent over the past two decades (Milanovic 2012). This represents a colossal concentration of the world's produced wealth, an active process of redistribution, if you will, from direct producers to those who hold concentrated financial power.

Despite their starkly unequal usage, the radical productivity of the socio-ecological forces of which we are part should provide a core substance for our visions of the future. Our conceptualisations of climatic change, therefore, need to be fundamentally intertwined with existing divisions of labour, control of productive assets and uneven distributions of consumption power. We cannot be certain of the climatic shifts that may occur over the coming decades. We can, however, know that we already collectively produce enough to address substantive needs and more across the global population. It is now time that we focus on how to organise these capacities according to a social logic grounded in collective justice and security. To do so in the current era of climatic transformation, we need urgently to work out how to produce ourselves differently.

References

Agarwal, B. 2014. "Food sovereignty, food security and democratic choice: critical contradictions, difficult conciliations." *Journal of Peasant Studies* forthcoming.

Akram-Lodhi, A. H. 2013. *Hungry for Change: Farmers, Food Justice and the Agrarian Question*. Halifax, NS: Fernwood.

Altieri, M., and P. Koohafkan. 2008. *Enduring Farms: Climate Change, Smallholders and Traditional Farming Communities*. Penang: Third World Network.

Altieri, M., and P. Roge. 2011. "The Ecological Role and Enhancement of Biodiversity in Agroecosystems." Pp. 15–32 in *Agriculture, Biodiversity and Markets: Livelihoods and Agroecology in Comparative Perspective*, edited by Stewart Lockie and David Carpenter. London: Earthscan.

Andreasson, S. 2005. "Accumulation and growth to what end?" *Capitalism, Nature, Socialism* 16(4): 57–76.

Bernstein, H. 2014. "Food sovereignty via the 'peasant way': a sceptical view." *Journal of Peasant Studies* forthcoming.

Borras, S., R. Hall, I. Scoones, B. White, and W. Wolford. 2011. "Towards a better understanding of global land grabbing: an editorial introduction." *Journal of Peasant Studies* 38(2): 209–16.

Buttel, F. 1997. "Some Observations on Agro-Food Change and the Future of Agricultural Sustainability Movements." Pp. 344–65 in *Globalising Food: Agrarian Questions and Global Restructuring*, edited by David Goodman and Michael Watts. London: Routledge.

De Schutter, O. 2010. *Report Submitted by the Special Rapporteur on the Right to Food*. Geneva: United Nations.

Foster, J. B. 1999. *The Vulnerable Planet: A Short Economic History of the Environment*. New York: Monthly Review Press.

Gliessman, S. R. 2007. *Agroecolog: The Ecology of Sustainable Food Systems*. Boca Raton, FL: CRC Press.

Holt-Gimenez, E., and A. Shattuck. 2011. "Food crises, food regimes and food movements: Rumblings of reform or tides of transformation?" *Journal of Peasant Studies* 38(1): 109–44.

Ingold, T. 2000. *The Perception of the Environment: Essays on Livelihood, Dwelling and Skill*. London: Routledge.

Lin, B., M. Jahi Chappell, J. Vandermeer, G. Smith, E. Quintero, R. Bezner-Kerr, . . . I. Perfecto. 2011. "Effects of industrial agriculture on climate change and the mitigation potential of small-scale agro-ecological farms." *CAB Reviews: Perspectives in Agriculture, Veterinary Science, Nutrition and Natural Resources* 6(20): 1–18.

Martínez-Alier, J. 2002. *The Environmentalism of the Poor: A Study of Ecological Conflicts and Valuation*. Cheltenham, UK: Edward Elgar.

Martínez-Alier, J. 2011. "The EROI of agriculture and its use by the Via Campesina." *Journal of Peasant Studies* 38(1): 145–60.

McMichael, P. 2013. *Food Regimes and Agrarian Questions*. Halifax, NS: Fernwood.

Milanovic, B. 2012. "Global Income Inequality by the Numbers: in History and Now." *World Bank Development Research Group: Policy Research Working Paper* 6259: 1–30.

Mitchell, T. 2002. *Rule of Experts: Egypt, Techno-Politics, Modernity*. Berkeley: University of California Press.

Nigam, A. 2011. *Desire Named Development*. New Delhi: Penguin.

OECD. 2008. *Measuring Sustainable Production*. New York: Organisation for Economic Co-operation and Development.

Patel, R. C. 2007. *Stuffed and Starved: Markets, Choice and the Battle for the World's Food System*. Toronto: HarperCollins.

Rice, J. 2007. "Ecological unequal exchange: Consumption, equity, and unsustainable structural relationships within the global economy." *International Journal of Comparative Sociology* 48(1): 43–72.

Rice, J. 2009. "North south relations and the ecological debt: Asserting a counter-hegemonic discourse." *Critical Sociology* 35(2): 225–52.

Rostow, W. W. 1960. *The Stages of Economic Growth: A Non-Communist Manifesto*. Cambridge: Cambridge University Press.

Schumacher, E. F. 1973. *Small Is Beautiful: A Study of Economics as if People Mattered*. London: Blond and Briggs.

Swyngedouw, E. 2010. "Apocalypse forever? Post-political populism and the spectre of climate change." *Theory, Culture and Society* 27(2–3): 213–32.

Taylor, M. 2009. "Displacing insecurity in a divided world: Global security, international development and the endless accumulation of capital." *Third World Quarterly* 30(1): 147–62.

UNCTAD. 2013. *Trade and Environment Review 2013*. Geneva: United Nations.

UNEP. 2012. *Avoiding Future Famines: Strengthening the Ecological Foundation of Food Security through Sustainable Food Systems*. Nairobi: United Nations Environment Programme.

UNEP. 2014. *Assessing Global Land Use: Balancing Consumption with Sustainable Supply*. Geneva: United Nations Environment Programme.

Watts, M. 1983. *Silent Violence: Food, Famine and Peasantry in Northern Nigeria*. Berkeley: University of California Press.

Weis, A. J. 2007. *The Global Food Economy: The Battle for the Future of Farming*. London: Zed Books.

Winson, A. 2013. *The Industrial Diet: The Degradation of Food and the Struggle for Healthy Eating*. Vancouver: UBC Press.

Woodward, D., and A. Sims. 2006. "Growth Is Failing the Poor: The Unbalanced Distribution of the Benefits and Costs of Global Economic Growth," DESA Working Paper #20. Geneva: United Nations.

World Bank. 2007. *World Development Report 2008: Agriculture for Development*. Oxford: Oxford University Press.

World Bank. 2009. *World Development Report 2009: Development and Climate Change*. Washington, DC: World Bank.

Index